OECD Studies on Water

Groundwater Allocation

MANAGING GROWING PRESSURES ON QUANTITY AND QUALITY

OECD

BETTER POLICIES FOR BETTER LIVES

Please cite this publication as:
OECD (2017), *Groundwater Allocation: Managing Growing Pressures on Quantity and Quality*, OECD Studies on Water, OECD Publishing, Paris.
http://dx.doi.org/10.1787/9789264281554-en

ISBN 978-92-64-28152-3 (print)
ISBN 978-92-64-28155-4 (PDF)
ISBN 978-92-64-28429-6 (epub)

Series: OECD Studies on Water
ISSN 2224-5073 (print)
ISSN 2224-5081 (online)

Photo credits: Cover © filippoph – Fotolia.com.

Corrigenda to OECD publications may be found on line at: *www.oecd.org/about/publishing/corrigenda.htm.*

Foreword

The intensifying competition for surface and groundwater resources is widely documented. The OECD Environmental Outlook to 2050 *highlights that water resources are already over-used or over-allocated in many places, with global demand expected to increase by 55% between 2000 and mid-century. Groundwater withdrawals have risen sharply over the past years – increasing nearly tenfold according to some estimates. These pressures, projected to be further exacerbated by climate change, have already made water allocation an urgent issue in a number of countries and one that is rising on the agenda in many others. Within this context, the OECD published the report* Water Resources Allocation: Sharing Risks and Opportunities *in 2015 to strengthen the evidence base and develop policy guidance to improve the design of allocation regimes.*

Building on this work, further analysis was undertaken to examine the specific challenges related to groundwater and how allocation arrangements can be designed in light of groundwater's distinctive features. The analysis builds on a series of case studies to document groundwater allocation challenges in a variety of contexts and provides policy guidance to assess and enhance allocation regimes for groundwater.

This report is an output of the OECD Environment Directorate. It was prepared by Kathleen Dominique and Marit Hjort, with guidance and input from Xavier Leflaive. The case studies on groundwater allocation were prepared by Marit Hjort, with input from delegates of the OECD Working Party on Biodiversity, Water and Ecosystems and in-country experts, notably: P.G. Pedersen, Chief advisor, Unit of Water Resources, Agency for Water and Nature Management, Ministry of Environment and Food, Denmark; Dr. Sharon Megdal, Director, Water Resources Research Center, University of Arizona; Jiro Hiratsuka, Ministry of Environment, Japan and experts from the local government in Kumamoto City; Daniel Rivera, International Cooperation Management, National Water Commission of Mexico; Ana Fueyo and Dr.Alberto López-Asenjo Garcia, Ministry of Agriculture, Spain, José Ángel Rodríguez-Cabellos, Head of the Planning Office in Guadiana River Authority, Spain; Larry French, Director of Groundwater, Rima Petrossian, Manager of Groundwater Technical Assistance, and Kimberly Friesen Leggett, Media Relations Specialist, Texas Water Development Board, C. E. Williams, General Manager, Panhandle Groundwater Conservation District; Jérémy Devaux, French Ministry of the Environment, Energy and the Sea, Jean-Daniel Rinaudo, Bureau de recherches géologiques et minières *(BRGM), France, Floriane Di Franco, Permanent Assembly of Agricultural Chambers, France, Patrice Garin, member of the research collective* Gestion de l'Eau, Acteurs, Usages, *France; Dr. R.C. Jain, Former Chairman, Central Ground Water Board and Central Ground Water Authority, Ministry of Water Resources, River Development and Ganga Rejuvenation, Government of India.*

The authors are also grateful to colleagues and experts who provided valuable comments on the report, including Simon Buckle, Guillaume Gruère, Paul O'Brien, Hannah Leckie (OECD Secretariat) and Ian Barker, of Water Policy International, Henry Leveson-Gower, DEFRA, UK and Professor Mike Young, University of Adelaide. Editorial support from Janine Treves and administrative support from Angèle N'Zinga are also gratefully acknowledged.

Table of contents

Tables

Figures

Follow OECD Publications on:

http://twitter.com/OECD_Pubs

http://www.facebook.com/OECDPublications

http://www.linkedin.com/groups/OECD-Publications-4645871

http://www.youtube.com/oecdilibrary

OECD Alerts http://www.oecd.org/oecddirect/

Acronyms

ALS	Abstraction Licensing Strategy
ADWR	Arizona Department of Water Resources
AWBA	Arizona Water Banking Authority
CAMS	Catchment Abstraction Management Strategies
CAP	Central Arizona Project
CONAGUA	National Water Commission of Mexico
COSTAS	Technical committees for groundwater, Mexico
CSWUA	Council for Sustainable Water Use in Agriculture
DFCs	Desired future conditions
EFI	Environmental flow indicator
EU	European Union
GCD	Groundwater conservation district
GDMP	Guadiana District Management Plan, Spain
GEB	Gujarat Electricity Board, India
GL	Giga litres
GWMU	Groundwater management unit
GRACE	NASA's Gravity Recovery and Climate Experiment
LEMA	*La Loi sur l'Eau et les Milieux Aquatiques*, France (Law on Water and Freshwater Ecosystems)
NGO	Non governmental organisation
NWRP	National Water Reserves Programme, Mexico
OECD	Organisation for Economic Co-operation and Development
OUGC	*organismes uniques de gestion collective*, France (single collective management bodies)
PES	Payment for ecosystem services
RBC	River basin council
RBO	River basin organisation
SGMA	Sustainable Groundwater Management Act of California
S&R	Storage and recovery
TEC	Kumamoto Technology Centre
TEV	Total economic value
TWDB	Texas Water Development Board
UGSP	Upper Guardiana Special Plan
UN	United Nations
U.S.	United States
WFD	EU Water Framework Directive
ZRE	*zone de repartition d'eau*, France (water distribution area)

Groundwater Allocation
Managing Growing Pressures on Quantity and Quality

Executive summary

As the predominant reservoir of freshwater on Earth, groundwater provides an important source of water supply for drinking, irrigation and industry and contributes to sustaining groundwater-dependent ecosystems, such as streams and wetlands. Pressures on the quantity and quality of the resource have increased significantly over recent decades. Globally, groundwater withdrawals have risen sharply; nearly tenfold in the past 50 years (Shah et al., 2007). At the same time, the resource is becoming increasingly degraded due to pollution and saline intrusion. Unsustainable groundwater use creates negative environmental externalities, including land subsidence, saline intrusion and the deterioration of groundwater-dependent ecosystems. Groundwater depletion also increases the cost of use, as pumping is required from ever-increasing depths, which may disadvantage small scale users. This depletion can also result in water shortage directly affecting users, with an impact on economic activities.

These mounting pressures have largely outpaced the modernisation of groundwater allocation regimes. Allocation regimes consist of the combination of policies, laws, regulations and institutional arrangements (entitlements, licenses, permits, etc.) that determine who is able to use water resources, how, when and where. In practice, many current groundwater allocation regimes are strongly conditioned by historical water usage patterns that evolved during periods when the resource was more abundant, demand was lower and access was minimally regulated or not at all. Acute governance challenges arise from the lack of data, fragmented legislation and the largely decentralised use of the resource. The entrenchment of weak or contradictory policies, such as under-pricing water or subsidising energy to pump groundwater, can make improving allocation arrangements contentious and costly. However, failure to improve allocation policies undermines the range of societal benefits from groundwater via extractive and non-extractive uses (e.g. for the environment) both today and in the future.

The benefits obtained from groundwater take many forms – from the economic value derived from productive uses for drinking water supply, industry and irrigation, to the ecological value provided by supporting key species in groundwater-dependent ecosystems to the option value of storing groundwater as a buffer against future water shortages. Groundwater allocation policies need to account for these different types of extractive and non-extractive values as well as balance the needs of current and future generations.

This report examines the distinctive features of groundwater and sets out policy guidance for groundwater allocation. This guidance should be used as a supplement to the general guidance on allocation in the *OECD Health Check for Water Resources Allocation*. The Health Check consists of a series of 14 questions ("checks") to identify whether key elements of an allocation regime are in place and how their performance could be improved. The full *Health Check* is set out in Chapter 2 of this report. Part II of this report analyses nine case studies (Denmark; Tucson, Arizona; Kumamoto, Japan; Mexico; the Upper Guadiana Basin,

Spain; Texas; France; Gujarat, India and North China), highlighting how elements of the Health Check can be addressed in diverse contexts.

A number of distinctive features of groundwater systems (compared to surface water) merit particular attention in the design of allocation regimes. There is significant scientific uncertainty about the state (quality and quantity) of groundwater resources and data on groundwater use are scarce and incomplete. There is a need to better understand how groundwater may be interconnected with surface water so as to manage the resources conjunctively, and monitor how groundwater use is changing over time. This requires an assessment of groundwater resources with a view to determining where abstraction may give rise to negative externalities.

To respond to the rapid growth of unregulated groundwater use, many governments have taken action to redefine groundwater ownership and use rights as within the public domain. This provides the basis for a legally enforceable regulatory regime. As the resource is increasingly brought under the public domain, a clear process for transferring from private ownership to regulated use needs to be put into place. Customary rights to access the resource also need to be considered.

Groundwater resources consist of both stocks and flows, which require a long-term exploitation strategy that considers both. Some aquifers are considered non-renewable (containing "fossil" groundwater), so the use of these resources is akin to irreversible mining. Only a portion of groundwater resources (consisting of total stocks and flows) should be considered as exploitable. From an economic perspective, optimal groundwater exploitation would maximise the present value of benefits minus costs, which can guide efforts to define an abstraction limit on the resource. Setting such a "cap" on abstraction requires balancing extractive and non-extractive uses uses (e.g. flows for ecosystem needs, protection of water quality) and current and future uses.

Groundwater generally exhibits the characteristics of a common pool resource, which makes excluding users from access difficult and costly. Users often access the resource directly, in a decentralised way. This makes monitoring groundwater use technically demanding and costly. New monitoring technologies, such as satellite-based telemetry, are showing promise in improving groundwater monitoring, however these still need to be complemented by ground-based measurements. When metering each user is not practicable or too costly, governments can consider using collective entitlements to allocate water to a group of users within a specific area.

Even a well-designed allocation regime can be undermined by perverse incentives in other sectors, such as subsidies that encourage over-consumption of groundwater or pollution that degrades water quality. Electricity or irrigation subsidies can encourage excessive groundwater pumping. Policies to safeguard groundwater quality by reducing potential contamination from pesticides, fertilisers, urban run-off and other pollution sources are particularly important.

As scarcity increases and the value of water use rises, the case for the introduction of a more elaborate allocation regime grows stronger. In the early stages of developing a resource, a relatively simple allocation regime can be used with decisions made conservatively to avoid over-allocation and depletion. However, the basic building blocks of a robust regime should still be put into place at an early stage to avoid lock-in to unsustainable use and allow for adjustment at least cost, as needed, over time. Adequate monitoring and analysis of water resources should be in place before problems become severe and allow policymakers to adjust the allocation regime as resource use intensifies. A periodic "health check" can provide a pragmatic approach to realise the benefits of improved allocation.

PART I

Key information and guidance for groundwater policy

Groundwater Allocation
Managing Growing Pressures on Quantity and Quality

PART I

Chapter 1

Overview of groundwater resources and prevailing trends

This chapter provides an overview of trends documenting increasing pressures on groundwater resources. It examines the range of benefits obtained from groundwater, including the economic value derived from productive uses, the ecological value provided to groundwater-dependent ecosystems, and the option value the resource provides as a buffer against future water shortages. The chapter then reviews the distinctive features of groundwater and their relevance for allocation policy design.

Introduction

Groundwater is a valuable natural resource, which provides an important source of water supply for drinking, irrigation and industry in many parts of the world and also contributes to sustaining groundwater-dependent ecosystems. Pressures on the quantity and quality of the resource have increased significantly. Globally, groundwater withdrawals have risen sharply and the resource is becoming increasingly degraded due to pollution and saline intrusion (Margat and van der Gun, 2013). However, groundwater allocation policies have generally not kept pace with these increasing pressures. There are inherent challenges involved in assessing the status of groundwater and investment in monitoring the resource has been inadequate to date (Foster et al., 2013). In many countries, there are persistent problems related to the efficient and equitable use of groundwater (GEF et al., 2015a), reducing the benefits that individuals and society reap from the resource, today and in the future.

Recent work by the OECD[1] and others[2] has contributed to improving policy guidance on water resources allocation, groundwater governance and managing groundwater in agriculture. However, some key gaps remain. In particular, guidance on how the various elements that comprise an allocation regime[3] (policies, laws, regulations and institutional arrangements) can be designed to accommodate the distinctive features of groundwater is lacking. Building on previous work, in particular the 2015 OECD report *Water Resources Allocation: Sharing Risks and Opportunities*, this report aims to fill this gap. Specifically, it focusses on how allocation regimes for groundwater or conjunctively managed surface and groundwater systems can be designed to bring about the desired policy outcomes, in terms of economic efficiency, environmental effectiveness and social equity.[4] Drawing on an assessment of groundwater's distinctive features and nine case studies of groundwater allocation in practice, the report provides guidance for designing policies that balance different types of extractive and non-extractive uses as well as the needs of current and future generations.

A valuable natural resource under increasing pressure

Groundwater systems make up the predominant reservoir and strategic reserve of freshwater on Earth[5] (Foster and Chilton, 2003). It provides a drinking water source for around half of the global population (Margat and van der Gun, 2013) and accounts for an increasing share for agricultural use making up around 40% of consumptive irrigation water, covering just under 40% of irrigated land globally (OECD, 2015a). More than 60% of abstracted groundwater is consumed by agriculture in arid and semi-arid regions, producing 40% of the world's food (Morris et al., 2003). Industrial uses (including mines and energy production) are also important, accounting for over one-fifth of total groundwater abstraction in some countries (Germany, Japan, Brazil, and the Philippines, among others) (Margat and van der Gun, 2013).

Groundwater and surface water systems are closely interlinked in most places on Earth and human activities, such as water abstraction, irrigation and artificial drainage, have intensified these interactions (GEF et al., 2015a). Often, a substantial portion of groundwater flow emerges to join surface water, supporting the base flow of surface water bodies (Margat and van der Gun, 2013). Also, groundwater withdrawals may be used as a substitute for surface water withdrawals, and vice versa. Thus, groundwater and surface water allocation need to be studied and managed conjunctively, not in isolation, where possible.

The intensifying use and competition for water resources is widely documented (WRI, 2016; OECD, 2012; UNESCO, 2012; Vörösmarty et al., 2010). The OECD *Environmental Outlook to 2050* highlights that water resources are already over-used or over-allocated in many places, with global demand expected to increase by 55% between 2000 and mid-century (OECD, 2012).

Generally, there is significant scientific uncertainty about the state (quality and quantity) of groundwater due in no small part to that fact that it is largely an "invisible" resource stored underground (discussed further below). Data on groundwater use are scarce and remain incomplete (Margat and van der Gun, 2013; Shah et al, 2007), yet some general trends are clear. Globally, groundwater withdrawals[6] have risen almost tenfold in the past 50 years (Shah et al., 2007). Between 1960 and 2000, the rate of groundwater depletion more than doubled (Wada et al., 2010). This boom in groundwater abstraction, driven by population growth and the associated increasing demands for water, food and income, has no precedent in history (Margat and van der Gun, 2013). Advances in drilling and pumping technology have lowered the cost of groundwater abstraction and contributed to greater exploitation of the resource. The rise in intensive use of groundwater by millions of small scale farmers is so striking that it has been dubbed a "silent revolution" (Llamas and Martínez-Santos, 2005).

During the second half of the 20th century, groundwater abstraction has followed a pattern similar to total water withdrawals. The most pronounced increases have been observed in countries where current groundwater withdrawals are the highest (Margat and van der Gun, 2013). Intensive groundwater withdrawal is particularly prevalent in countries such as Israel, Mexico, Spain, Turkey, the southwest of the United States (U.S.), Bangladesh, northern parts of The People's Republic of China (hereafter "China"), northern India, Indonesia, Iran, Pakistan and Saudi Arabia (Margat and van der Gun, 2013; Shah et al., 2007). Notably there is a correlation between high aridity, groundwater dependence and abstraction intensity. Figure 1.1. illustrates the top ten groundwater abstracting countries (in terms of volume of water abstracted).

The total global withdrawal of groundwater was estimated at 8% of the mean global groundwater renewal in 2010, but this is highly variable among countries and may reach up to 50% in some cases (GEF, et al., 2015a). Groundwater abstraction is projected to stablise or slightly decrease in industrialised countries, while abstraction is projected to continue to increase in countries where economic and demographic growth is substantial and where irrigation is a significant user, such as countries in Asia (Margat and van der Gun, 2013).

Groundwater is also becoming increasingly degraded due to pollution and saline intrusion (GEF, et al., 2015a; Margat and van der Gun, 2013). This degradation can be caused by the introduction of contaminants, such as those in fertilisers or pesticides, or by changes in the groundwater regime (often triggered by increasing withdrawals), which may increase saline intrusion or the concentration of existing contaminants, such as arsenic (Margat and van der Gun, 2013). Both current pollution as well as legacy pollution are problematic. Land use changes, such as extending impermeable surfaces in urban areas, can reduce groundwater recharge and contribute to pollution. Agricultural intensification can increase

Figure 1.1. **Top ten groundwater abstracting countries**

Abstraction as of 2010

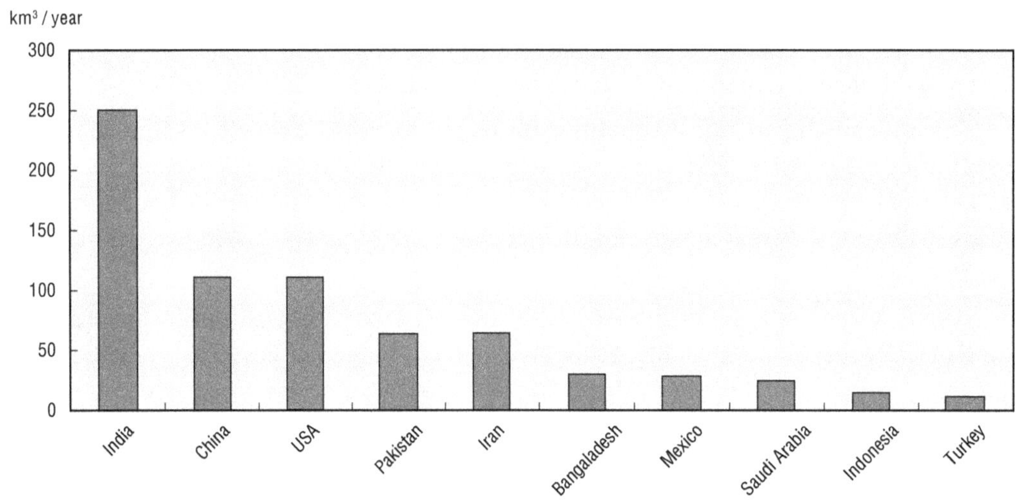

Source: Based on data from Margat and van der Gun, 2013.

diffuse pollution and leaching of contaminants, such as nitrates, into groundwater (GEF et al., 2015a; Margat and van der Gun, 2013). At the same time, excessive recharge due to leaking public water supply networks, for example, can cause salinisation, alkalinisation and waterlogging (Margat and van der Gun, 2013). In the energy sector, the recent shale gas boom has increased pressure on the resource in some regions and provoked public concern over potential groundwater contamination risks (e.g. the leakage of fracturing fluids, hydrocarbons or saline water) (IEA, 2012).[7] Degraded groundwater quality reduces its suitability for drinking (and other uses that require high quality water), increases the cost of treatment and can exacerbate water scarcity where degraded groundwater quality limits use.

Climate change is projected to reduce renewable surface water and groundwater resources in some regions, further intensifying competition for water (IPCC, 2014). Climate change is driving an intensification of the water cycle (Huntington, 2006), changing precipitation patterns, increasing evapotranspiration, impacting groundwater recharge and water quality, as well as increasing the frequency and intensity of extreme events (Bates et al., 2008; IPCC, 2014). In addition, sea level rise due to climate change contributes to saline intrusion in coastal aquifers (IPCC, 2014; Clifton et al, 2010). Higher water demand due to increasing temperatures and greater variability in precipitation (inter-annual and seasonal changes) is expected to particularly affect areas where mean groundwater recharge is expected to decrease (Margat and van der Gun, 2013).[8] Climate change is also expected to greatly expand groundwater's role in meeting water demand in some regions (Margat and van der Gun, 2013; OECD, 2015a).

Unsustainable groundwater use creates negative environmental externalities, including saline intrusion, land subsidence and reduction in spring flows and base flow, which puts stress on groundwater-dependent ecosystems such as wetlands (Box 1.1). This undermines the values (economic, environmental, social and cultural[9]) supported by groundwater resources and can result in irreversible damage (Margat and van der Gun, 2013; GEF et al., 2015a). Many cities are affected by land subsidence due to groundwater depletion, such as Tokyo, Shanghai, Calcutta, Venice, Mexico City and San Francisco (GEF et al., 2015a).

Box 1.1. **Groundwater-dependent ecosystems**

Ecosystems that rely on a supply of groundwater to function are considered to be groundwater-dependent ecosystems. They include rivers, lakes, riparian habitats, wetlands, springs, subterrean aquifers as well as estuarine and nearshore marine ecosystems. These ecosystems provide important ecosystems services, including food production, water purification, recreation, as well as habitats for migratory birds or rare plant and invertebrate species. Groundwater supports these ecosystem services through the provision of water (some ecosystems are fully dependent on groundwater), nutrients, buoyancy (as in the case of peatland bogs) and stability of water temperature. The reliance of ecosystems on groundwater may be continuous or periodic (seasonal or only during a limited period every few years).

While the contribution of groundwater to these ecosystems is recognised as vital, there are numerous, complex interactions, which are still poorly understood. Further, there is scant evidence about how groundwater depletion, pollution and land use change affects groundwater-dependent ecosystems. To adequately protect these ecosystems, more study is needed on their status, how they function and the impacts of land and water use, pollution and climate change.

Source: Kløve et al. (2011a); Kløve et al. (2011b).

Groundwater depletion also increases the cost of use, as pumping is required from ever-increasing depths, which may put small scale users at a disadvantage in terms of access to the resource (OECD, 2015a). The increasing cost of use may or may not be directly borne by groundwater users, depending on whether the cost of electricity or fuel to operate pumps is subsidised.[10] This depletion can result in water shortage directly affecting users and can have indirect impacts on economic activities, such as lost earnings and foregone profits (OECD, 2013). Groundwater depletion could become the greatest threat to urban water supplies in several regions in the coming decades (OECD, 2012), resulting in potential high replacement costs to secure alternative sources of water.

The benefits of groundwater: Estimating value

Groundwater resources serve multiple purposes and provide value to individuals, ecosystems, farms, firms, and society (including indigenous communities) in various ways. The benefits obtained from groundwater take many forms – from the economic value derived from productive uses for drinking water, industry and irrigation to the ecological value provided by supporting groundwater-dependent ecosystems to the option value of storing groundwater to use as a buffer against future water shortages. How much groundwater is left in aquifers and how much is abstracted for various uses; who is able to use these resources, how, when and where are questions that directly affect the benefits that individuals and society obtain from groundwater today and in the future. These questions are determined by allocation regimes, whether formal or informal.

The valuation of groundwater in alternative direct uses and also *in situ* non-extractive uses can provide important information to policy makers seeking to design allocation regimes that maximise the benefits of groundwater. Estimating the value of groundwater is a technically complex challenge.[11] However, the total economic value (TEV) approach provides a useful conceptual framework that can be used to identify the various ways in which groundwater generates benefits. The concept consists of several distinct types of values: 1) use values, 2) option values, and 3) non-use, or "passive" values. The use value reflects the direct use of the resource, such as groundwater abstracted for drinking water as well as non-extractive (indirect use) value, which derives from the ecosystem services the resource provides. These ecosystem services include base flow for streams and rivers, which supports recreational uses (fishing, boating) and hydropower production, among others. Figure 1.2 illustrates the various components of TEV and how they relate to groundwater.

Figure 1.2. **Total Economic Value of groundwater**

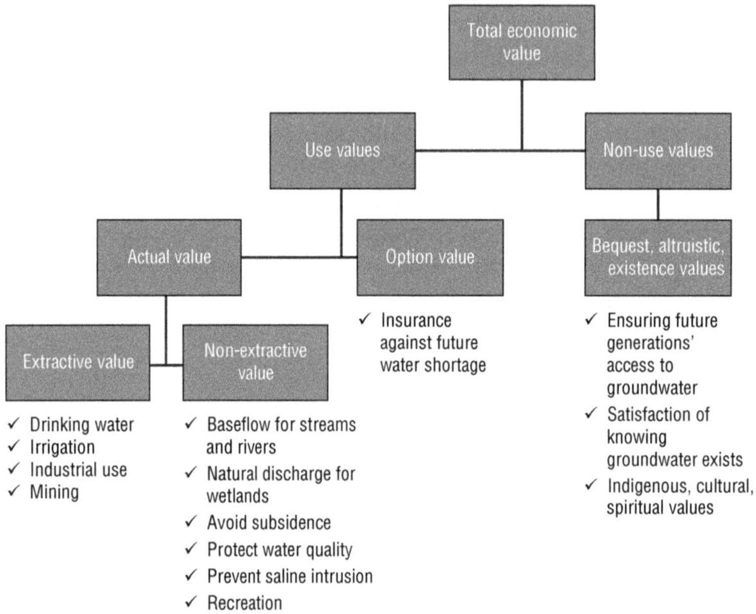

Source: Author, adapted from Qureshi et al. (2012); Johns and Ozdemiroglu (2007).

Groundwater allocation policies need to account for different types of extractive and non-extractive values as well as balance the needs of current and future generations. Allocating groundwater for non-extractive uses that leave groundwater *in situ* as well as option values and bequest values often require trade-offs with current extractive uses (Qureshi et al., 2012). The benefits from direct groundwater use vary considerably by type of use. For example, drinking water is a high value use, from an economic and social perspective. The value added per unit of water use by industry is typically higher than use for irrigation[12] (GEF et al., 2015a). Non-extractive values can be considerable, such as when groundwater supports vital ecosystem services or protects water quality. A range of valuation methods can be used to estimate these values, including revealed preference methods (such as actual or simulated markets, travel cost, hedonic property values, avoidance expenditures) and stated preference methods (such as contingent valuation or choice experiments)[13] (Tientenberg and Lewis, 2016). Overall, the economic value of groundwater varies greatly over time and space, depending on the quality, reliability and degree of substitutability of the resource and how it generates benefits. Box 1.2 provides an illustration of the economic value of consumptive groundwater use in Australia.

Groundwater resources can also be considered as natural capital, generating income flows through direct uses and sustaining ecosystem services through indirect uses. From this perspective, the level of the "stock" of groundwater is vital to the generation of such

Box 1.2. **Consumptive groundwater use in Australia: A valuable contribution to the economy**

A recent study aggregates disparate estimates of the value of consumptive groundwater use in Australia. Overall, an estimated 3 500 giga litres (GL) of groundwater provides a direct use value of between AUD 1.8 to 7.2 billion per year, with a midpoint estimate of AUD 4.1 billion. As for groundwater's contribution to Australia's GDP, the estimated range spans from AUD 3 to 11 billion per year, with a midpoint estimate of AUD 6.8 billion. This is in addition to AUD 419 million of use value annually to households. These estimates are only a partial view of the total economic value of groundwater, as non-extractive uses and options values have not been quantified.

The table below summarises the breakdown by groundwater use by sector.

Table 1.1. **Estimated value of consumptive groundwater use in Australia**

Sector	Groundwater volumes (ML)	Direct value-add (AUD millions)	Direct value add (AUD/ ML) range and central estimate	Contribution to GDP (AUD millions)
Agriculture – irrigation	2 050 634	$410	$30-500 $200	$820
Agriculture – drinking water for livestock	–	$393		$818
Mining	410 615	$1 129	$500-5 000 $2 750	$1 637
Urban water supply	303 230	$606	$1 000-3 000 $2 000	$1 146
Households	167 638	$419	$1 400-6 400 $2 500	n/a
Manufacturing and other industries	588 726	$1 177	$1 000-3 000 $2 000	$2 355
Total	**3 520 843**	**$4 136**		**$6 777**

Note: Figures provided are broad estimates using data from a range of sources between the years 2006 and 2012.
Source: Adapted from Deloitte Access Economics, 2013.

flows both today and in the future (GEF et al., 2015a). Box 1.3 provides an illustration of the estimated value of groundwater in the Kansas High Plains Aquifer in the U.S. using a natural capital approach.

Box 1.3. **Groundwater as natural capital: The Kansas High Plains Aquifer**

Fenichel et al. (2016) developed a framework to assess natural capital asset prices consistent with economic capital theory and applied it to the Kansas High Plains Aquifer. This aquifer supports significant food production in the U.S., but is rapidly depleting. The analysis shows that between 1996 and 2005, the profits attributable to the Kansas portion of the aquifer dropped from USD 2.3 billion to USD 1.2 billion. This amounted to a loss of approximately USD 110 million per year (2005 USD, 3% discount rate) of capital value due to groundwater withdrawal and changes in aquifer management.

By way of illustration, the study highlights that this yearly decline in wealth is twice as large as the state's investment in school infrastructure (an investment in physical capital, which enables the development of human capital) over the period.

Source: Fenichel et al., 2016.

A need for robust groundwater allocation regimes

Without effective policies to control abstraction, there is little or no incentive for users to limit groundwater pumping and conserve the resource, resulting in an inefficient allocation of the resource. Further, perverse incentives, such as subsidies for electricity to pump groundwater, can exacerbate pressure on the resource.

A 2014 OECD survey of water resources allocation practices in OECD and select non-OECD countries[14] confirmed that in most countries, water allocation arrangements are strongly conditioned by historical preferences and usage patterns, locking in water use to uses that may no longer be as valuable today as they were years ago (OECD, 2015b). Moreover, certain water uses may no longer be viable in the future, considering the potential magnitude of some climate change scenarios.

Weak allocation policies may be a particular problem for groundwater. Historically, water legislation has focused on surface water resources, while groundwater legislation has lagged behind remaining fragmented, incoherent or simply ignored in many countries (Mechlem, 2012). More prevelant private ownership of groundwater (as compared to surface water) can limit the authority of governments to control abstraction (Figure 1.3) (OECD, 2015b). The rule of capture, whereby farmers have the right to access and use any groundwater under their land, is still dominant in some places (GEF et al., 2015a; OECD, 2015a).

Figure 1.3. **Public and private ownership of ground and surface water resources**

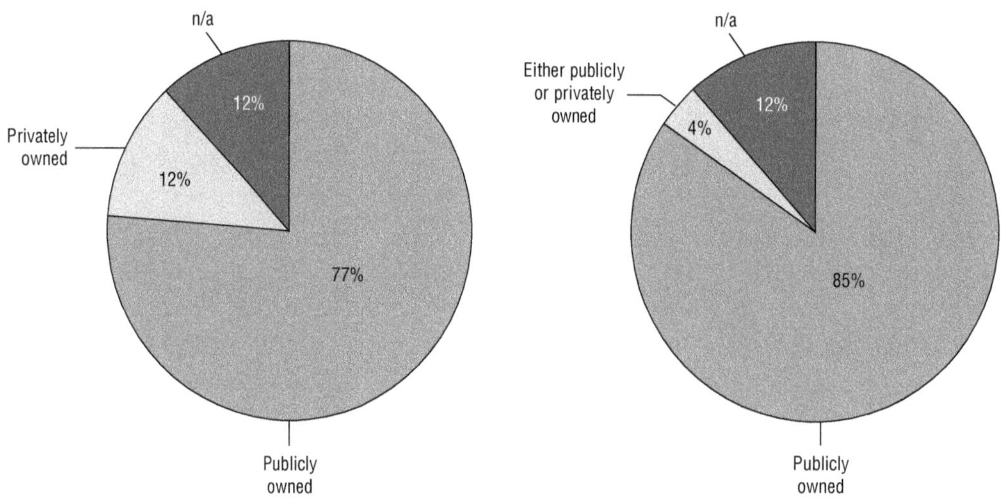

Note: It is important to note that "ownership" here refers to ownership of the resource itself, not the entitlement or right to use the resource. Does not include Switzerland. "n/a" refers to cases where water resources are not subject to legal ownership (either public or private). In these cases, water resources may be designated as *res nullis*, or "ownerless property" in legal terms.

Source: OECD (2015b), *Water Resources Allocation: Sharing Risks and Opportunities*, OECD Studies on Water, OECD Publishing, Paris, *http://dx.doi.org/10.1787/9789264229631-en*. See country profiles at *www.oecd.org/env/resources/water-resources-allocation.htm*.

The rapid growth of unregulated groundwater use has spurred many countries to try to redefine groundwater ownership and use rights as within the public domain and to support this with a legally enforceable regulatory regime (GEF et al, 2015a). Even where groundwater is formally declared by law as a public good and users only have usufructuary rights (or "use" rights), the perception that the resource is still private property can linger on (Mechlem, 2012).

Distinctive features of groundwater and implications for policy design

Given the close interlinkages between groundwater and surface water in most parts of the world, ground and surface water allocation need to be managed conjunctively wherever possible. Still, there are a number of distinctive features of groundwater systems (as compared to surface water), which deserve consideration in the design of allocation regimes. This section provides a brief summary of these features.

Uncertainty about state and use of the resource

There is greater scientific uncertainty about the state (quality and quantity) of groundwater resources as compared to surface water, due in no small part to that fact that it is an "invisible" resource stored underground. Although flows are generally easier to measure than stocks, recharge measurements are very difficult (OECD, 2015a). Data on groundwater use are scarce and remain incomplete (Margat and van der Gun, 2013; Shah et al., 2007). Shallow aquifers have been inventoried globally, but comprehensive mapping and assessment of larger, deeper aquifers has typically only been undertaken in developed countries (GEF et al, 2015a).

Groundwater is often available to multiple users without visible control or monitoring. Monitoring aquifers is technically demanding and costly, leaving the scientific understanding of many aquifers incomplete and complicating groundwater management (Mechlem, 2012). Relative to surface water, groundwater is much more poorly monitored and well metering requirements are only a recent development in many countries (Wheeler et al., 2016). However, well metering and reporting is on the rise in a growing number of groundwater management areas.[15] In some basins, remote telemetry may be used to monitor groundwater use (Aladjem and Sunding, 2015). For example, NASA's Gravity Recovery and Climate Experiment (GRACE) is the first satellite mission of its kind to map surface and groundwater resources and changes in these resources over time. It does so by monitoring changes in the Earth's gravitational field with an unprecedented temporal and spatial resolution and precision (NASA, 2016).[16]

Physical characteristics: Stocks, flows and quality

The quantity of groundwater resources can be characterised by two key variables: stock (volume stored) and flow (rate of renewal). With the exception of "fossil" groundwater, for most aquifers, the flow is a more relevant variable for characterising groundwater quantities than the stock (Margat and van der Gun, 2013). All aquifers have natural inflows and outflows of water, but the rates and speed of recharge and discharge vary greatly. In general, the storage capacity of aquifers is high relative to inflows (Giordano, 2009). It can take up to decades before groundwater depletion manifests as lower pressure in wells or lower water tables.

Whereas the time between surface water leaving and entering the system may be a matter of weeks, for groundwater, it can take up to thousands of years (Oki and Kanae, 2006). Thus, groundwater can serve as a strategic reserve and buffer against shocks (GEF et al., 2015b) and provide an important contribution to resilient water management. Conjunctive management of a range of water sources (aquifers, rivers, reservoirs, treated wastewater or desalination) gives water managers and users a portfolio of options, which generally provides more secure, flexible and resilient supplies.

Some aquifers are considered "non-renewable" since the time it takes to renew them can extend to hundreds of millennia. Most "pure non-renewable" aquifers (containing "fossil" groundwater) are located in North Africa and the Arabian Peninsula. They are large, deep, confined aquifers that formed long ago and receive an insignificant amount of recharge. The use of these groundwater resources may be likened to irreversible mining[17] (OECD, 2015a).

Groundwater recharge can be increased by inefficient water systems (e.g. leaky irrigation systems). Thus, unless the water allocation regime properly accounts for return flows (the residual of water abstracted, but not consumed), improvements in efficiency of use can result in unintended consequences (such as reduced groundwater recharge) (OECD, 2015b).

The quality of groundwater is generally superior to that of surface water (especially with regards to bacterial contamination), hence its importance as a source of drinking water. However, groundwater is particularly vulnerable to long-term, cumulative pollution, which may only manifest after significant time lags (Margat and van der Gun, 2013).

Compared to surface water, groundwater is relatively insulated from the direct effects of climate variability and climate change. However, as noted above, the impacts of climate change on groundwater systems are expected to be considerable. Higher water demand due to rising temperatures, greater variability in precipitation (inter-annual and seasonal changes) as well as an increasing risk of drought is expected to greatly expand groundwater's role in meeting water demand (OECD, 2015a; Margat and van der Gun, 2013).

A need for a long term exploitation strategy considering both stocks and flows

Only a portion of groundwater resources (consisting of total stocks and flows) should be considered as exploitable. To limit negative externalities, groundwater exploitation may be subject to significant constraints in order to maintain ecosystem services, avoid land subsidence or quality degradation. Further, the exploitation of some groundwater bodies or a portion of groundwater may be technically infeasible or economically undesirable, when pumping costs outweigh benefits. Exploitable groundwater resources can be augmented, typically via artificial recharge or through induced recharge (Margat and van der Gun, 2013).

Determining how to allocate groundwater stocks and flows among current and future users is a critical element of an allocation regime. Several long term exploitation strategies can be employed to determine the appropriate level of abstraction over time: 1) a sustainable yield[18] strategy aims to abstract inflows and keep the groundwater flow in a balanced state. This strategy aims to harvest inflows sustainably (limiting abstraction to the portion of recharge or inflow that is not needed to sustain base flows); 2) a mixed strategy, with depletion during a limited period and abstraction at a sustainable rate in the longer term and possibly recharge to help the stock recover; and 3) a mining strategy whereby stocks are progressively depleted (Margat and van der Gun, 2013). Figure 1.4 illustrates these three exploitation strategies.

Any of these strategies may be deliberately chosen, or in many cases, may become a *de facto* strategy in an unregulated situation or one where attempts are made to control pumping and let stocks recover. Box 1.4 provides an illustration of how "sustainable management" is defined in the 2014 Sustainable Groundwater Management Act (SGMA), a major reform in California.

From an economic perspective, optimal groundwater exploitation would maximise the present value of benefits minus costs (Qureshi et al., 2012). An efficient allocation of the resource requires that the marginal benefit (or value) of extracting an additional unit of water at all times and locations equals the full marginal opportunity cost of extracting that unit of water. The latter consists of the actual marginal costs of extracting a unit of water

Figure 1.4. **Groundwater exploitation strategies**

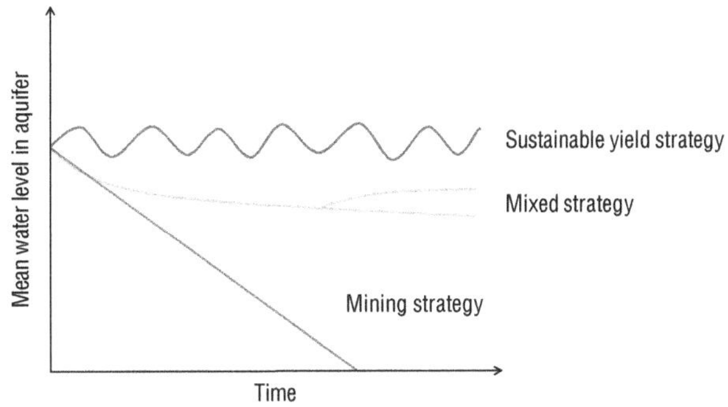

Source: Adapted from Margat and van der Gun, 2013 and BGS, 2009.

Box 1.4. **Defining a "sustainable" groundwater management strategy**

California depends on groundwater for a significant portion of its water supply (40% in an average year, even more in drier years). Until recently, groundwater use was largely unregulated, contributing to substantial depletion. In 2014, the state passed Sustainable Groundwater Management Act (SGMA), which came into effect on 1 January 2015. For the first time in California's history, this major reform empowers local authorities to adopt and enforce groundwater management plans to put resource use on a sustainable footing.

According to SGMA, sustainable groundwater management is defined as "the management and use of groundwater in a manner that can be maintained during the planning and implementation horizon without causing undesirable results". These "undesirable" effects include:

- Chronic lowering of groundwater levels, but excluding reductions in groundwater levels during a drought if they are offset by increases in groundwater levels during other periods;
- Significant and unreasonable reductions in groundwater storage;
- Significant and unreasonable seawater intrusion;
- Significant and unreasonable degradation of water quality;
- Significant and unreasonable land subsidence; and
- Surface water depletions that have significant and unreasonable adverse impacts on beneficial uses.

Over-drafted basins are required to achieve groundwater sustainability by 2040 or 2042, depending on the completion of management plans. The State Water Resources Control Board has the authority to intervene if deadlines are not met and establish an interim plan.

Source: Water Education Foundation (2015).

in addition to the present value of the increase in future marginal costs resulting from the absence of that unit of water[19] (Qureshi et al., 2012).

Non-renewable groundwater resources represent a special case. In this case, the stock is the focus of the exploitation strategy, rather than the flow. The main constraint on mining non-renewable resources is the rising cost of extraction due to declining water levels. Allocation policies have to balance the benefits of current abstraction and future abstraction and should account for the scarcity rent of exploiting a non-renewable resource.

Often a common pool resource, difficult to exclude users from access

Groundwater often exhibits the characteristics of a common pool resource, in terms of high rivalry and low excludability, although this is not always the case. As Brozovic et al. (2006) demonstrate, the impact of one user on others varies depending on the hydrological conditions of the aquifer. When the amount of water released in an aquifer due to a reduction in pressure (storativity) is low and the speed of lateral flow (transmissivity) is high, groundwater can easily flow across the aquifer. Thus, the effect of a user's pumping is widely transmitted through the aquifer. On the other hand, an aquifer with high storativity and low tranmissivity is closer to a private good than common pool resource (Huang et al., 2012).

The degree of connection with surface water systems can also affect whether groundwater is characterised as a private good or common pool resource (OECD, 2015a), which has important implications for how groundwater should be managed (OECD, 2015b; Huang et al., 2012). Considering the specific collective action problems posed by common pool resources (Ostrom, 1990), this is particularly relevant for allocation policies, including how to appropriately define water entitlements and determining rules regarding water trading (if permitted) (Wheeler et al., 2016).

Decentralised access by users on demand

As aquifers can cover large areas spread out horizontally, users can directly access water on demand under their land and, in the case of shallow aquifers, at relatively low cost (OECD, 2015a). Thus, the access and use of groundwater is usually more decentralised than for surface water, and does not always require co-operation among users, as each operators controls his or her own pumps (OECD, 2015a). Since groundwater is more poorly monitored than surface water and well metering requirements are only a recent development in many countries, groundwater markets may be more difficult to establish than surface water markets (Box 1.5).

The cost of accessing groundwater is usually borne by the user. It consists mainly of a fixed cost for a well and a variable cost for pumping, which depends on the state of the resource and the cost of energy (Garrido et al., 2006). In the case of surface water, the fixed cost, related to infrastructure to store and transport the water, is often borne by public agencies. For both surface and groundwater resources, the variable costs include abstraction charges if they are in place, which usually do not reflect actual costs (OECD, 2015a). Energy consumption for groundwater pumping is on the rise and appears to be a significant share of total energy consumption in countries where groundwater is intensively exploited (such as India, China and the U.S.), although lack of adequate data prohibits reliable estimates (Margat and van der Gun, 2013). Subsidies for energy to pump groundwater (such as in India or Mexico) provide a perverse incentive to over-exploit the resource. In addition, groundwater pumping can generate multiple external costs, including falling water levels, wetland degradation and land subsidence (Margat and van der Gun, 2013).

Acute governance challenges due to fragmented legislation, decentralised use and lack of data

Groundwater governance faces many of the same challenges as surface water governance, but at times to a greater degree. For example, challenges for both surface water and groundwater governance arise from the mismatch between administrative boundaries and the

Box 1.5. **Groundwater markets: Challenges and opportunities**

Groundwater markets are less common than surface water markets, but have emerged in a number of countries, including Australia, China, India, the US, Oman, Pakistan. In principle, markets can improve the efficiency of allocation by shifting water use to higher value uses. The functioning of groundwater markets differs depending on the context. In China, Oman or India, groundwater is usually sold and transported to be used on another property. With informal groundwater markets, such as in India or Pakistan, farmers who can afford large wells and pumps sell water to smaller farmers who cannot afford such infrastructure in exchange for labour or cash (Olmstead, 2010). In Australia and the US, trading usually involves selling water entitlements to another user within the same aquifer. However, in the US some major transfers involve purchasing water from farms and pumping it to distant cities (for example in Arizona, California or Texas) (Wheeler et al., 2016). In addition, groundwater banking schemes can be used to transfer water among users and shift use over time. Groundwater banking consists of storing surface water in aquifers during abundant periods for use during drier periods. This is a relatively cost-effective means to increase water supply during droughts and offset loss of seasonal storage historically provided by snowpack-fed systems (Wheeler et al., 2016).

Groundwater markets face distinct challenges, including accounting for the characteristics of the aquifer, uncertainties about the resource and aquifer boundaries, changes in water quality and local drawdown impacts (Wheeler et al., 2016). Groundwater trading can change the location of pumping and thus, the distribution and magnitude of pumping externalities (Aladjem and Sunding, 2015). To address this issue, zoning schemes (such as in the Murray Darling Basin in Australia) may be used or trading ratios (such as in Nebraska, US), which adjust for the different impacts of a change in pumping location (Aladjem and Sunding, 2015).

Establishing formal groundwater markets entails adequately defining water entitlements, establishing and enforcing a regulatory framework and accounting for resource costs and externalities (GEF et al, 2015a). Accurate monitoring and measurements of groundwater use is a prerequisite for the establishment of a well-functioning market. A growing number of groundwater management areas require well metering and reporting (Aladjem and Sunding, 2015).

Source: Wheeler et al., 2016; Aladjem and Sunding, 2015; GEF et al., 2015a; Olmstead, 2010.

relevant scale for water governance, typically river basins. In addition, aquifer boundaries generally do not correspond to river basins, which compounds these governance challenges.

In most parts of the world, groundwater governance is generally poor or absent (GEF et al., 2015a). Historically, groundwater legislation has lagged behind legislation for surface water. While legislation on groundwater is found in nearly all countries, it is often fragmented, incoherent or outdated (GEF et al 2015a; Mechlem, 2012). Groundwater legislation is usually comprised of rules on ownership, abstraction and use based on entitlements, protection from pollution, and assignment of roles and responsibilities to competent authorities. Laws and enforcement responsibilities related to quality are often distinct from other aspects of groundwater management (GEF et al., 2015a). Land law has important implications for access to groundwater and its protection (Mechlem, 2012), whereas for surface water land law relates mainly to riparian rights.

Both the effective management of surface water and groundwater may be undermined by a lack of coherence among sectoral policies. In the case of groundwater, subsidies for energy used to pump groundwater can be particularly problematic. Both surface water and groundwater are typically managed in a decentralised way, but since access to groundwater tends to be more decentralised than access to surface water, it does not always require co-operation among users. This lack of co-ordination among groundwater users can cause significant issues in circumstances where use of the resource widely affects availability for other users.

The lack of data and knowledge of groundwater resources and limited monitoring systems contributes to weak governance. The issue or unregulated wells is prevalent in some regions, such as Southern Europe (OECD, 2010). In European Mediterranean countries, as many as half of the wells may be unregistered or illegal (EASAC, 2010). Further, as an "invisible" resource the lack of awareness of the state of groundwater resources hinders stakeholder engagement. In general, the state of groundwater governance varies widely and is closely linked to the stage of development of the resource and the level of development of the country (GEF et al. 2015a).

Conclusion

Groundwater is under increasing pressure due to intensive abstraction and degraded quality, which reduces the value of the resource and the ecosystem services it provides as well as increases pumping and treatment costs and other negative effects, such as land subsidence. Groundwater is a valuable resource, providing benefits through direct productive uses, such as drinking water or irrigation, and indirect uses, such as flows for ecosystems. The resource also provides an option value, in that it can provide a buffer against future shortages and other values, such as ensuring availability for use by future generations.

Given the close interlinkages between surface and groundwater in many places, allocation need to be studied and managed conjunctively, not in isolation. However, there are a number of distinctive features of groundwater that require specific attention in allocation policy design. This includes the significant scientific uncertainty about the state (quality and quantity) of groundwater resources and scarce data on use. Since groundwater generally exhibits the characteristics of a common pool resource, excluding users from access can be difficult and costly. Acute governance challenges arise from the lack of data, fragmented legislation and largely decentralised use of the resource. Groundwater resources consist of both stocks and flows, which require a long-term exploitation strategy that considers both variables. The trend towards redefining ownership and use rights previously considered private property as within the public domain is a positive step towards encouraging more sustainable use, however evidence suggests that enforcement of laws and regulations on groundwater remains generally weak.

The following chapter sets out policy guidance for groundwater allocation in the form of a "Health Check", which can be used to assess the current state of allocation practice and identify areas for improvement. Part II of this report examines nine case studies (Denmark; Tucson, Arizona; Kumamoto, Japan; Mexico; the Upper Guadiana Basin, Spain; Texas; France; Gujarat, India and North China) to examine how various groundwater allocation challenges are being addressed in diverse contexts.

Notes

1. For example, the OECD report *Water Resources Allocation: Sharing Risks and Opportunities* (2015) provides a comprehensive analysis of water allocation policies in OECD and key partner countries and developed related policy guidance. The OECD report Drying Wells, Rising Stakes (2015) provides a comprehensive analysis of the economics and policies for groundwater management in agriculture in OECD countries.
2. A major, multi-year initiative on groundwater governance was recently completed by the Global Environment Facility, the United Nation's Food and Agriculture Organisation, UNESCO's International Hydrological Programme, the International Association of Hydrologists and the World Bank. The main project outcome, the Global Framework for Action provides a set of guidelines for groundwater governance at the local and national levels.
3. See the glossary defining key terms appended at the end of this report.
4. A framework detailing these elements at a general level and how they may influence policy objectives is set out in the report Water Resources Allocation: Sharing Risks and Opportunities (OECD, 2015b).
5. Groundwater accounts for around 30% of global freshwater and as much as 98% if water frozen in the polar ice caps and glaciers are excluded
6. Estimates of the proportion of abstracted groundwater that is actually consumed are scarce, with the exception of irrigation, which is on average around 80% (varying depending on overall irrigation efficiency). For domestic and industrial uses, the portion consumed is usually much smaller, but varies considerably (Margat and van der Gun, 2013).
7. The U.S. EPA has studied the link between fracking and drinking water in the U.S. and identified factors that are more likely to result in more frequent or severe impacts on drinking water resources. These include fracking in areas with low water availability (especially areas with limited or declining groundwater); spills of fracking fluids; inadequate wells; discharge of inadequately treated fluids, etc) (U.S. EPA, 2016).
8. This could cause severe problems, especially in small and shallow alluvial aquifers in arid and semi-arid regions (Van der Gun, 2009).
9. Cultural values of groundwater include indigenous values.
10. See case study on Gujarat, India, for an example. Also, in Mexico, Tarifa 9, is a preferential tariff for electricity to pump groundwater for rural users, which has led to overexploitation of many aquifers in water scarce regions (OECD, 2013 MRHMEX).
11. Given the limitations in understanding of all of the benefits of groundwater (environmental and otherwise) and the methodological challenges related to the economic evaluation of these benefits, the TEV approach does not provide an exhaustive view of all of the benefits of groundwater.
12. While the value added of food production may be lower than industry, in some regions, groundwater supports also supports other policy objectives, such as food security.
13. The appropriate method will vary, depending on the situation, the availability of state and the type of value being assessed. These methods present a number of challenges and typically only provide a partial estimation of values. However even a partial estimate can be preferable to ignoring such values entirely.
14. The survey collected information about 37 examples of allocation regimes in 27 OECD countries as well as Brazil, China, Colombia, Costa Rica, Peru and South Africa (OECD, 2015b).
15. For example, well metering is required in certain basins in the U.S. and can also be found in Australia, New Zealand and China (Aladjem and Sunding, 2015).
16. The decade-long study has documented that 21 of the world's 37 largest aquifers are being depleted. This novel approach is helping to fill gaps in the scarce data on freshwater resources, especially groundwater, but many findings are only relevant for very large aquifers.
17. Groundwater mining is geographically concentrated in four countries Saudi Arabia, Algeria, Libya and United Arab Emirates, which account for 86% of the total estimated global groundwater mining (Margat and van der Gun, 2013).
18. Sustainable yield is defined as the flux of groundwater that can be withdrawn from an aquifer without causing undesirable side effects, in particular without causing a permanent state of imbalance in the hydrological budget of an aquifer. It includes economic and environmental criteria and underlies the concept of "overexploitation" (Margat and van der Gun, 2013).

19. The increase in future marginal cost consists of: (1) the future increase in marginal costs of all extractors and (2) the marginal reduction of future non-extractive benefits that depend on water stock or flows from that water stock (Qureshi, 2012).

References

Aladjem and Sunding (2015), "Marketing the Sustainable Groundwater Management Act: Applying economics to solve California's groundwater problems", *Natural Resources and Environment*, Vol. 30(2), American Bar Association.

Bates, B. et al. (eds.) (2008), "Climate change and water", *Technical Paper VI*, Intergovernmental Panel on Climate Change (IPCC), UNFCCC Secretariat, Geneva.

Brozovi, N. et al. (2006), "Optimal management of groundwater over space and time", in Goetz, R.U. and D. Berga (eds.), *Frontiers in Water Resource Economics*, Springer, New York.

British Geological Survey (BGS) (2009), "Groundwater Information Sheet", London, *www.bgs.ac.uk/*.

Clifton, C. (2010), "Water and climate change: Impacts on groundwater resources and adaptation options", *Water Working Notes*, Note No. 25, June 2010, World Bank Group.

Deloitte Access Economics (2013), "Economic value of groundwater in Australia", National Centre for Groundwater Research and Training, Deloitte.

European Academies Science Advisory Council (EASAC) (2010), "Groundwater in the southern member states of the European Union", EASAC Policy Report 12, EASAC, Halle.

Fenichel, E. et al. (2016), "Measuring the value of groundwater and other forms of natural capital", *Proceedings of the National Academy of Sciences*, *www.pnas.org/cgi/doi/10.1073/pnas.1513779113*.

Foster, S. et al. (2013), "Groundwater: A global focus on the 'local resource'", *Current Opinion in Environmental Sustainability*, (5) 685-695.

Foster, S.S.D. and P.J. Chilton (2003), "Groundwater: The processes and global significance of aquifer degradation", *Philosophical Transactions of the Royal Society of London*, Series B-Biological Sciences 358, no. 1440: 1957-1972.

Garrido, A. et al. (2006), "Groundwater irrigation and its implications for water policy in semiarid countries: the Spanish experience", *Hydrogeology Journal*, Vol. 14(3), pp. 340-349.

Global Environment Facility (GEF) et al. (2015a), "Global Diagnostic on Groundwater Governance", GEF, United Nations Educational, Scientific and Cultural Organisation International Hydrological Programme (UNESCO-IHP), United Nationals Food and Agriculture Organisation (FAO), the World Bank Group (WB), and the International Association of Hydrologists (IAH), Special Edition for World Water Foum 7, 10 March 2015.

GEF et al. (2015b), "Global Framework for Action to Achieve the Vision on Groundwater Governance", GEF, FAO, UNESCO-IHP, IAH, and the World Bank, Special Edition for World Water Foum 7, 10 March 2015, *www.groundwatergovernance.org/fileadmin/user_upload/groundwatergovernance/docs/general/GWG_FRAMEWORK.pdf* (accessed 11 February 2016).

Giordano, M. (2009), "Global groundwater? Issues and solutions", *Annual Review of Environment and Resources*, Vol. 34(1), pp. 152-178.

Huang, Q. et al. (2012), "The effects of well management and the nature of the aquifer on groundwater resources", *American Journal of Agricultural Economics*, Vol. 95(1), pp. 94-116.

Huntington, T. (2006), "Evidence for intensification of the globalwater cycle: Review and synthesis", *Journal of Hydrology*, Vol. 319, Elsevier, pp. 83-95, *http://dx.doi.org/10.1016/j.jhydrol.2005.07.003*.

IEA (2012), *World Energy Outlook 2012*, OECD Publishing, Paris, *http://dx.doi.org/10.1787/weo-2012-en*.

Intergovernmental Panel on Climate Change (IPPC) (2014), *Climate Change 2014 Synthesis Report Summary for Policymakers*", Intergovernmental Panel on Climate Change, *www.ipcc.ch/pdf/assessment-report/ar5/syr/AR5_SYR_FINAL_SPM.pdf* (Accessed 22 February 2016).

Johns, H. and E. Ozdemiroglu (2007), "Assessing the value of groundwater", Environment Agency, UK, *www.aueb.gr/users/koundouri/resees/uploads/Econ%20Val%20GW.pdf* (Accessed 10 August 2016).

Kløve, B. et al. (2011a), "Groundwater dependent ecosystems. Part I: Hydroecological status and trends", *Environmental Science and Policy*, vol. 14, pp. 770-781, *http://dx.doi.org/10.1016/j.envsci.2011.04.002*.

Kløve, B. et al. (2011b), "Groundwater dependent ecosystems. Part II: Ecosystem services and management in Europe under risk of climate change and land use intensification", *Environmental Science and Policy*, vol. 14, pp. 782-793, *http://dx.doi.org/10.1016/j.envsci.2011.04.005*.

Llamas, M.R. and P. Martínez-Santos (2005), "Intensive groundwater use: Silent revolution and potential source of social conflicts", *Journal of Water Resources Planning and Management*, September-October 2005, pp. 337-341.

Margat, J. and J. van der Gun (2013), *Groundwater around the World: A Geographic Synopsis*, CRC Press/ Balkema, Taylor and Francis, London.

Mechlem, K. (2012), "Legal and institutional frameworks", *Thematic Paper* 6, Groundwater Governance: A Global Framework for Action, GEF, FAO, UNESCO-IHP, IAH, and the World Bank.

Morris, B.L. et al. (2003), "Groundwater and its susceptibility to degradation: A global assessment of the problem of options for management", *Early Warning and Assessment Report Series*, RS. 03-3, UNEP, Nairobi, Kenya.

NASA (2016), "GRACE tellus gravity recovery and climate experiment", *http://grace.jpl.nasa.gov/* (accessed 29 November 2016).

OECD (2015a), *Drying Wells, Rising Stakes: Towards Sustainable Agricultural Groundwater Use*, OECD Studies on Water, OECD Publishing, Paris, *http://dx.doi.org/10.1787/9789264238701-en*.

OECD (2015b), *Water Resources Allocation: Sharing Risks and Opportunities*, OECD Studies on Water, OECD Publishing, Paris, *http://dx.doi.org/10.1787/9789264229631-en*.

OECD (2013), *Water Security for Better Lives*, OECD Studies on Water, OECD Publishing, Paris, *http://dx.doi.org/10.1787/9789264202405-en*.

OECD (2012), *OECD Environmental Outlook to 2050: The Consequences of Inaction*, OECD Publishing, Paris, *http://dx.doi.org/10.1787/9789264122246-en*.

OECD (2010), *Sustainable Management of Water Resources in Agriculture*, OECD, Paris, France, *http://dx.doi.org/10.1787/9789264083578-en*.

Oki, T. and S. Kanae (2006), "Global hydrological cycles and world water resources", *Science*, Vol. 313, No. 5790.

Olmstead, S. (2010), "The economics of managing scarce water resources", *Review of Environmental Economics and Policy*, Vol. 4(2), pp. 179-198, *http://dx.doi.org/10.1093/reep/req004*.

Ostrom, E. (1990), *Governing the Commons: The Evolution of Institutions for Collective Action*, Cambridge University Press.

Qureshi, M.E. et al (2012), "Factors determining the economic value of groundwater", *Hydrogeology Journal*, vol. 20: 821-829.

Saak, A.E. and J.M. Peterson (2007), "Groundwater use under incomplete information", *Journal of Environmental Economics and Management*, 54(2):214-228.

Shah, T. et al. (2007), *Groundwater: A Global Assessment of Scale and Significance*, International Water Management Institute (IWMI), 2007.

Tietenberg, T. and L. Lewis (2016), *Environmental and Natural Resource Economics*, 10th Edition, Routledge, London and New York.

UNESCO (2012), "Managing water under uncertainty and risk", *The United Nations World Water Development Report* 4, United National Educational Scientific and Cultural Organisation, Paris.

U.S. Environmental Protection Agency (EPA) (2016), "Hydraulic fracturing for oil and gas: Impacts from the hydraulic fracturing water cycle on drinking water resources in the United States (Final Report)", U.S. Environmental Protection Agency, Washington, DC, EPA/600/R-16/236F.

Van der Gun (2009), "Climate change and alluvial aquifers in arid regions: examples from Yemen", Chapter 11 in: Ludwig et al. (eds): *Climate Change Adaptation in the Water Sector*, Co-operative Programme on Water and Climate, Earthscan, London, pp. 159–176.

Vörösmarty, C.J. et al. (2010), "Global threats to human water security and river biodiversity", *Nature*, Vol. 467, 30 September, *http://dx.doi.org/10.1038/nature09440*.

Wada, Y. et al. (2010), "A worldwide view of groundwater depletion", *Geophysical Research Letters*, 37, L20402, *http://dx.doi.org/10.1029/2010GL044571*.

Water Education Foundation (2015), *The 2014 Sustainable Groundwater Management Act: A Handbook to Understanding and Implementing the Law*, Water Education Foundation, *www.watereducation.org/sites/main/files/file-attachments/groundwatermgthandbook_oct2015.pdf* (accessed 27 November 2016).

Wheeler, S.A. et al (2016), "Lessons to be learned from groundwater trading in Australia and the United States", *Integrated Groundwater Management*, A. Jakeman (ed.), *http://dx.doi.org/10.1007/978-3-319-23576-9_20*.

World Resources Institute (WRI) (2016), *Aqueduct: Measuring and Mapping Water Risk*, World Resources Institute, *www.wri.org/our-work/project/aqueduct* (accessed 4 April 2016).

Groundwater Allocation
Managing Growing Pressures on Quantity and Quality

PART I

Chapter 2

The OECD health check for water resources allocation: Groundwater guidance

This chapter sets out policy guidance for groundwater allocation. The guidance in this chapter should be used as a supplement to the OECD Health Check for Water Resources *when assessing allocation arrangements of groundwater systems and in cases where surface and groundwater systems are managed conjunctively. The guidance in this chapter first reiterates some of the general principles that apply broadly to all allocation regimes, then describes how the specific features of groundwater can be considered.*

Many current groundwater allocation regimes are strongly conditioned by historical water usage patterns that evolved during periods when access to the resource was minimally regulated or not at all. Thus, they are typically not well-equipped to deal with growing pressures on the resource (OECD, 2015a). The entrenchment of weak or contradictory policies, such as under-pricing water or subsidising energy to pump groundwater, can make improving allocation arrangements contentious and costly. However, failure to improve allocation policies undermines the range of benefits that society could reap from groundwater via extractive and non-extractive uses (e.g. for the environment) both today and in the future. Building on the analysis in the previous chapter, this chapter sets out tailored guidelines for good practice for allocation in settings where groundwater is either the main source of water supply or managed conjunctively with surface water resources.

The policy guidance set out below builds on the general framework and guidance in the *OECD Health Check for Water Resources Allocation* in the 2015 OECD publication *Water Resources Allocation: Sharing Risks and Opportunities*. The guidance specific to groundwater described in this chapter should be used as a supplement to the *OECD Health Check for Water Resources*. The guidance first reiterates some of the general principles that apply broadly to all allocation regimes, then describes how the specific features of groundwater should be considered. Box 2.1 provides a summary of the Health Check. Each of the elements are discussed in detail in this chapter.

The *Health Check* uses a series of questions ("checks") to identify whether key elements of an allocation regime are in place and how their performance could be improved. In some cases, options for the design of elements are proposed. References are also made to the case studies in Part II of this report. The case studies provide illustrations of how a particular "check" is applied in practice or highlight some of the challenges faced in implementation.

The policy guidance in the *Health Check* provides general principles, but these need to be tailored to specific conditions to ensure the allocation arrangements are "fit for purpose". There is a wide variety of groundwater systems, which can be categorised by their geological features (sand and gravel; sandstone; karst; volcanic; or basement aquifers) or by the hydrogeological setting (major aquifers; complex hydrogeological structures; shallow aquifers). Groundwater systems can also be characterised according to socio-economic features (degree, intensity and type of use) or their stage of development (OECD, 2015b). Most groundwater systems interact with surface water systems to some extent, calling for conjunctive management that accounts for complex interactions between the resources. These factors should be considered in the assessment and design of allocation arrangements.

As groundwater systems are more intensively used, the benefits of a more elaborate allocation regime increase. In the early stages of developing a groundwater resource, a relatively simple allocation regime can be used with decisions made conservatively to avoid over-allocation and depletion. However, the basic building blocks of a robust regime should still be put into place at an early stage to avoid lock-in to unsustainable use and allow for adjustment at least cost, as needed, over time. As scarcity increases and the value of water use rises, the case for the introduction of a more elaborate allocation regime increases. When water over-allocation or unsustainable use already exists, there is an opportunity to use the characteristics of a more elaborate allocation regime to reduce the extent of the problem and bring use in line with sustainable limits (OECD, 2015a). Adequate monitoring and analysis of groundwater resources should be in place before problems become severe and allow policymakers to adjust the allocation regime as resource use intensifies.

Box 2.1. **OECD Health Check for Water Resources Allocation**

Check 1. Are there accountability mechanisms in place for the management of groundwater allocation that are effective at the aquifer or other relevant scale?

Check 2. Is there a clear legal status for all water resources (surface and groundwater, as well as alternative sources of supply)?

Check 3. Is the availability of water resources (surface and groundwater, as well as alternative sources of supply) and possible scarcity well-understood?

Check 4. Is there an abstraction limit ("cap") that reflects *in situ* requirements and sustainable use?

Check 5. Is there an effective approach to enable efficient and fair management of the risk of shortage that ensures water for essential uses?

Check 6. Are there adequate arrangements in place for dealing with exceptional circumstances (such as a drought or severe pollution events)?

Check 7. Is there a process for dealing with new entrants and for increasing or varying existing entitlements?

Check 8. Are there effective mechanisms for monitoring and enforcement, with clear and legally robust sanctions?

Check 9. Are water infrastructures in place in order for the allocation regime to function effectively?

Check 10. Is there policy coherence across sectors that affect water resources allocation?

Check 11. Is there a clear legal definition of water entitlements?

Check 12. Are appropriate abstraction charges in place for all users that reflect the impact of the abstraction on resource availability for other users and the environment?

Check 13. Are obligations related to return flows and discharges properly specified and enforced?

Check 14. Does the system allow water users to reallocate water among themselves to improve the allocative efficiency of the regime?

Check 1. Are there accountability mechanisms in place for the management of groundwater allocation that are effective at the aquifer or other relevant scale?

In general, authorities and organisations responsible for allocation should have well-defined roles and accountability mechanisms that actually work in practice, as well as sufficient resources (financial and otherwise) to execute their functions. A clear and **transparent process** should be in place to facilitate stakeholder engagement in the determination of a sustainable exploitation strategy and other key allocation decisions (see the case study of Tucson, Arizona, Part II).

In the case of groundwater, a **Management Plan** (or other similar planning instrument) addressing groundwater resources or conjunctively managed surface and groundwater resources that has the status of a statutory instrument that must be followed can be used to set out a clear framework for allocation. The plan should have clear and explicit links to groundwater protection policies to ensure quality and quantity issues are managed in a co-ordinated way. For example, in the Murray-Darling Basin, Australia, the Basin Plan limits water use at environmentally sustainable levels by determining long-term sustainable diversion limits for both surface and groundwater resources. A key component of the Basin Plan is the environmental watering plan, which co-ordinates all environmental watering across the Basin. The Plan also contains a water quality and salinity management plan and water quality targets which influence how environmental flows and the water resources are managed (OECD, 2015a). The Catchment Management Strategies used in England are illustrated in Box 2.2 (see also the case studies of Texas, U.S. and France, Part II).

Box 2.2. **Integrated management of surface and groundwater resources in England**

In England, the Environment Agency has established a comprehensive network of river flow measurement sites and groundwater level monitoring boreholes, together with widespread ecological monitoring. It uses a Resource Assessment Methodology to determine how much water is reliably available for abstraction on a catchment by catchment basis. By taking account of the volume of water already allocated for abstraction, and how much (in terms of flow or level) the environment needs, it can then set out how much water is potentially available for further abstraction. It can also identify where unsustainable abstraction is taking place and the solutions to address the problem. This information is published in Catchment Abstraction Management Strategies (CAMS) which cover every catchment in England.

The detailed Abstraction Licensing Strategies (ALS) which make up the CAMS integrate groundwater availability and river flows, so that the effects of groundwater abstraction on surface water features are a key part of the assessment. An Environmental Flow Indicator (EFI) is used to assess whether river flows are sufficient to support a healthy ecology, and the EFIs control the volume available for abstraction across the entire flow range. The aim is not just to protect low flows, but to maintain flow variability. The more sensitive that the river's ecology is to changes in flow or level, the more restrictive the conditions on abstraction. The impact of groundwater abstraction on river flows is considered in the management strategy for each aquifer unit, together with the sensitivity of wetland features to changes in groundwater level.

All ALS are consulted on and published, so that the availability of resources and the conditions on new abstractions (where allowed) are clearly set out. Each Groundwater Management Unit (GWMU) will fall into one of three categories (listed in the table below):

Table 2.1. **Groundwater licensing dependent on status of the resource**

GWMU resource availability category	Implication for abstraction licensing
Water available for licensing	Groundwater unit balance shows groundwater available for licensing. New licenses can be considered depending on impacts on other abstractors and on surface water.
Restricted water available for licensing	Groundwater unit balance shows more water is licensed than the amount available, but that recent actual abstractions are lower than the amount available OR that there are known local impacts likely to occur on dependent wetlands, groundwater levels or cause saline intrusion. In restricted units, no new consumptive licences will be granted. Water may be available through licence trading.
Water not available for licensing	Groundwater unit balance shows more water has been abstracted based on recent amounts than the amount available. No further consumptive licences will be granted.

Source: Adapted from UK Environment Agency (2016), "Managing Water Abstraction", Environment Agency, Bristol.

Box 2.2. **Integrated management of surface and groundwater resources in England** *(cont.)*

The intensive use of groundwater for public supply and irrigation has had an adverse effect on river flows and ecology across much of the south and east of England. In those parts of England where groundwater licences are still being issued many licences will contain a Groundwater Level Condition, which requires abstraction to reduce or stop once levels fall below a critical limit.

Source: Ian Barker (2016), personal communication; UK Environment Agency (2016).

Check 2. Is there a clear legal status for all water resources (surface and groundwater, as well as alternative sources of supply)?

A **clear legal status** should be in place for all types of water resources (surface and groundwater, as well as alternative sources of sypply, such as treated wastewater). This status needs to define whether the resources are publicly or privately owned, or in cases where there is no ownership of water resources, *per se*, who has the authority to determine access to the resource (see the case study of France, Part II). Any contradictory and overlapping legal arrangements relating to the ownership of the resource itself as well as legal entitlement to access and use water resources should be clarified.

In the case of groundwater, as the resource is increasingly brought under the public domain, a clear process for transferring from private ownership to regulated use should be put into place (see the case study of the Upper Guadiana Basin, Spain, Part II). Customary rights to access the resource also need to be considered.

Check 3. Is the availability of water resources (surface and groundwater, as well as alternative sources of supply) and possible scarcity well-understood?

A **robust scientific basis** is needed to identify the available water resources. In the case of groundwater, there is a need to understand how groundwater may be interconnected with surface water, and how groundwater use is changing over time. This requires an assessment of groundwater resources with a view to determining where abstraction may give rise to negative externalities (see examples in the case studies of Denmark and Mexico, Part II). The comprehensiveness of the assessment should correspond to the degree of unsustainable use and/or quality degradation, with groundwater resources under greater pressure deserving more in-depth assessment as compared to those where depletion or quality degradation is not yet an issue and not expected to be in the near future. While new monitoring technologies, such as satellite-based telemetry,[1] are improving the scientific understanding of groundwater, in general, it is not possible, nor is it necessary, to obtain complete knowledge of water systems. Instead, the aim should be to acquire sufficient knowledge of the available groundwater resources in order to make appropriate and tailored decisions. The information should be made publically available in a way that encourages public understanding.

Managing **system interconnectivity** is essential for ensuring the hydrological integrity of the system. For instance, careful consideration needs to be given to the impact of groundwater bores located next to a river. In such situations, extraction from the bore may in fact actually be extraction from a river which has high connectivity to the groundwater resource. In order to avoid double-counting that will result in over-use in such circumstances, the amount of water that may be taken from the river needs to decrease and the amount of water taken from the aquifer can increase. Arrangements need to be in place to adjust for changes in flows between groundwater and surface water systems. Where economically viable, surface water or treated wastewater can be used for artificial or induced recharge (see examples in the case studies of Tucson, Arizona and Kumamoto, Japan, Part II). For example, water banking schemes that store water in aquifers to save for drier periods can be a relatively cost-effective way to increase water supplies during drought (Wheeler et al., 2016).

Check 4. Is there an abstraction limit ("cap") that reflects in situ requirements and sustainable use?

Defining a limit on the maximum volume or proportion of water that can be abstracted from a system is arguably the most difficult and yet most important challenges in managing water scarcity. Two types of abstraction limits are needed:

- A **long-term limit** that defines the maximum volume of water that can be abstracted at any point in time. Once this limit has been fully allocated, no new entitlements should be issued unless the process is accompanied by an arrangement that reduces someone else's entitlement by an equivalent amount. A mechanism to adjust the long-term limit is needed for adaptive management. This is especially the case in regions where recharge can be highly variable and the expectations of adverse impacts of climate change, but can also relate to the need to adapt to other drivers of change, for instance as a result of new scientific evidence about ecosystem needs. The long-term limit can be used to guide strategic water-dependent investments.
- A **short-term limit** on the amount of water that can be taken during a particular period. In addition to limits on the maximum amount of water that can be taken over the long-term, in most systems, it is also necessary to be able to adjust the amount of water that can be taken within a given time period, which can be adjusted depending on variations in recharge.

In the case of groundwater, only a portion of groundwater resources (consisting of total stocks and flows) should be considered as exploitable. Setting an abstraction limit requires consideration of the amount of water that should be left in the aquifer to meet non-extractive uses (e.g. flows for ecosystem needs, protection of water quality) and future uses (see examples in the case studies of Denmark, Mexico, Texas and France, Part II). From an economic perspective, optimal groundwater exploitation would maximise the present value of benefits minus costs, which requires balancing extractive and non-extractive uses and current and future uses.

Both policy-related and technical limitations on the quantity of groundwater available for sustainable use need to be recognised. Technical restrictions limit economically viable abstraction, when pumping costs exceed benefits. Policy-related limitations may include obligations related to minimum flows to ensure base flow of connected surface water bodies, environmental flows, or maintaining the groundwater stock to avoid subsidence or quality degradation. Environmental demands on groundwater can be significant and need to be accounted for in groundwater management plans. For example, in the Netherlands, environmental functions adjusted to different types of land use pose a major constraint on groundwater abstraction in Dutch groundwater policies and plans. In Spain, environmental needs are explicitly accounted for in the National Hydrological Plan (Margat and van der Gun, 2013).

Check 5. Is there an effective approach to enable efficient and fair management of the risk of shortage that ensures water for essential uses?

"Essential", high value water uses, such as drinking water, should be defined and assured the **highest priority** in case that temporary bans on water use are put into place. Water needs for the environment should also be secured (see the case study of Mexico, Part II).

In cases where surface and groundwater resources interact or can be used as substitutes, securing access for essential uses should be managed conjunctively.

Check 6. Are there adequate arrangements in place for dealing with exceptional circumstances (such as a drought or severe pollution events)?

The conditions that constitute an "exceptional circumstance", such as a drought or severe pollution event, need to be clearly specified (see the case study of France, Part II). Stakeholders should be involved in the process of determining what constitutes exceptional circumstances. A **responsible authority** that has authority to declare an exceptional circumstance and manage the response needs to be designated. Water users need to be informed regularly about the developments relating to exceptional circumstances and how they will be affected by the response. The more advance warning that users can be provided, the more opportunity that they will have to adjust their behaviour and effectively manage their risk.

In the case of groundwater systems, users may increase their reliance on groundwater pumping when a drought limits the availability of surface water. A severe pollution event may impact either surface water or groundwater supplies. Groundwater systems are particularly vulnerable to pollution, which can accumulate over long periods of time.

Check 7. Is there a process for dealing with new entrants and for increasing or varying existing entitlements?

When the defined resource pool is fully allocated, the resource should be considered "closed". Once access to the resource is closed, the only way a new entrant may secure an interest in abstracting water from the resource or an existing use may expand an existing entitlement is to ensure that **another user foregoes use of an equivalent amount**, thereby transferring the water entitlement to the new entrant or the existing user expanding an entitlement (see case studies of Upper Guadiana Basin, Spain and France, Part II). This applies generally to both surface and groundwater systems,

Check 8. Are there effective mechanisms for monitoring and enforcement, with clear and legally robust sanctions?

A robust allocation regime should aim for an appropriate level of **monitoring** of the resource, ecosystem requirements, abstractions, and recharge that reflects the level of pressure on the water resource (see case studies of Denmark and Texas, Part II). Rigorous monitoring requires monitoring the volume of water being taken by each user. This requires the installation of meters, meter reading, and accounting protocols. Appropriate accounting arrangements that track water use and consumption, as well as leases and trades (where permitted), need to be in place to support the monitoring of resource use and water entitlements. Appropriate **sanctions**, such as fines or curtailment of water entitlements need to be in place and applied as required (see case study of Texas, Part II).

Uncontrolled uses and any significant interception need to be periodically reviewed to gauge their potential impact on the integrity of the system. When uncontrolled uses and significant interceptions begin to have a significant impact on the water system, they must be brought into the formal water entitlement system. This sends a clear signal to existing entitlement holders that the expansion of these uses will not undermine the efficiency of any investments they have made.

In the case of groundwater, data on resource use are scarce and remain incomplete and monitoring aquifers is technically demanding and costly. New monitoring technologies, such as satellite-based telemetry, are showing promise in improving groundwater monitoring, however these still need to be complemented by ground-based measurements. When metering each user to not practicable or too costly, other options could be considered, such as monitoring a group of users with a collective entitlement within a specific area.

Check 9. Are water infrastructures in place in order for the allocation regime to function effectively?

Adequate water **infrastructures** are needed to store, treat, and deliver water to various users. A lack of such infrastructure can place constraints on the flexibility of allocation regimes. Authorities need to ensure that sustainable financing mechanisms are in place to support investment in water infrastructures and their operation and maintence.

While most groundwater users abstract water directly with their own pumps, water infrastructures may be needed to treat and deliver water to various uses and users and support artificial recharge for water banking or other purposes (see case study of Tucson, Arizona, Part II).

Check 10. Is there policy coherence across sectors that affect water resources allocation?

The existing policy settings related to water resources management as well as water-related sectors, such as agriculture, energy, and urban development need to be coherent (see case studies of Kumamoto, Japan and Texas, Part II).

In the case of groundwater, even a well-designed allocation regime can be **undermined by perverse incentives** in other sectors, such as subsidies that encourage over-consumption of groundwater or pollution that degrades water quality. For example, electricity or irrigation subsidies can encourage excessive groundwater pumping (see case studies of Mexico and Gujarat, India, Part II). Policies to protect groundwater quality by reducing potential contamination from pesticides, fertilisers, urban run-off and other pollution sources (such as a pollution tax), are particularly important (see case study of Denmark, Part II).

Check 11. Is there a clear legal definition of water entitlements?

Well-functioning allocation regimes need to have **clear, quantified, legally defined water entitlements,** whether formal legal entitlements or customary rights (see case studies of Denmark and Upper Guadiana Basin, Spain, Part II).

In the case of groundwater, options for defining how users can access water and how much they are allowed to take range from a requirement that the user own land above the groundwater to a requirement that all abstractions require water entitlements that are controlled and metered. Permits for wells or boreholes should require a pumping test to demonstrate the yield (to ensure that the groundwater resource can support the abstraction permitted in the associated entitlement) and any local effects, such as adversely affecting nearby boreholes, environmental uses (streams, wetlands), or the quality of the resource.

To improve the flexibility of the allocation regime, water entitlements can be **unbundled** from land titles, although, to date, this is not a widespread practice. There are benefits to defining water entitlements as a **proportion**, or shares, of the available resource pool (as opposed to an absolute volume) (see case study of Texas, Part II). This approach allows for flexibility to respond to changing conditions (e.g. increased or decreased recharge) without having to pay compensation for adjusting water entitlements. This approach is also consistent with the full assignment of risk. Conversion from a volumetric or seniority regime to a proportional regime is possible, although it may be challenging.

Water entitlements must be defined for an **appropriate duration**, with a clear, reasonable expectation for renewal (see case study of Denmark, Part II). This could be a fixed period of time, or water entitlements could be defined in perpetuity. The longer the entitlement is granted for, the more it will encourage long-term investment in water-related activities. Uses that require significant investment to benefit from the water entitlement merit a longer duration.

In the case of groundwater, the degree of uncertainty related to the availability of the resource and the potential adverse effects of abstraction should also inform the duration of the entitlement. A higher level of uncertainty about resource availability would justify a shorter duration to allow for further monitoring of the resource.

Check 12. Are appropriate abstraction charges in place for all users that reflect the impact of the abstraction on resource availability for other users and the environment?

Appropriate **abstraction charges** should be levied on users, in line with the "beneficiary pays" principle (see case study on Denmark[2]). An abstraction charge can provide an incentive to allocate and use water more efficiently, although its impact on behaviour will depend on the price elasticity of demand (where this is low, users are less responsive to price changes; where it is high, users are more responsive to price changes).

In the case of groundwater, in practice it is easier to apply an abstraction charge to large-scale uses, such as municipal water supply, industrial users, or large-scale irrigators, but is much more difficult and costly to apply to small-scale irrigators.

In designing an abstraction charge, the charge should reflect environmental and resource costs associated with abstraction. It should also contribute to recovering costs associated with managing the resource, including monitoring costs, which can be significant. The level of the abstraction charge can be differentiated based on the local context, to reflect differences in the vulnerability of the environment and other users to changes in the groundwater level (Ambec et al., 2016). When groundwater is metered, a volumetric charge can be applied. If groundwater use is unmetered, a flat abstraction charge or one based on a proxy, such as area of irrigated land, can be used as a more rudimentary alternative (Ambec et al., 2016). The charge should be set in a manner coherent with abstraction charges for surface water bodies, to account for potential substitution effects.

In the case of **non-renewable groundwater** resources, setting an efficient price, in theory, requires including Hotelling rents,[3] to reflect the trade-off between mining water now or in the future (Olmstead, 2010).

Check 13. Are obligations related to return flows and discharges properly specified and enforced?

Water entitlements need to be specified in a way that defines the **"net" amount of water** consumed, rather than the "gross" amount of water abstracted, when relevant. This requires accounting for water that has been abstracted, but returns a portion to the system via recharge. Notably, improvements in irrigation efficiency can significantly reduce groundwater recharge. In practice, there are numerous technical challenges that make it difficult (if not impossible) to measure net consumptive use with precision. However, rules of thumb can be applied to provide an estimation of net consumption according to the type of use. This approach can be used to maintain the integrity of the allocation regime, even while efficiency of use increases.

In considering the impact of return flows on groundwater, it important to recognise that recharge will be affected by a number of factors, including aquifer characteristics (e.g. unconfined shallow aquifers versus deep confined aquifers).

Check 14. Does the system allow water users to reallocate water among themselves to improve the allocative efficiency of the regime?

Once the elements of a robust allocation regime are in place, allowing water entitlement holders to **trade, lease or transfer** water entitlements can improve efficiency in allocation and resource use (see case studies of France, Gujarat, India and North China, Part II). To avoid potentially negative impacts of trading arising from changing the location of water use, water entitlements and trading arrangements must be consistent with the overall limits of the resource. Where the trade, lease or transfer of water entitlements is possible, clear rules should be in place to facilitate transactions. Voluntary forfeiture of un-used water entitlements should be provided for.

Transaction costs related to trading, leasing or transferring water entitlement and allocations should be kept as low as possible. This requires limiting trading costs to administrative costs that are unavoidable and also limiting third party interference in individual transactions.

Notes

1. See for example, reference to NASA's Gravity Recovery and Climate Experiment (GRACE) in Chapter 1.
2. Abstraction charges for groundwater are present in a number of countries, including, but not limited to: Australia, Belgium, the Czech Republic, Denmark, Estonia, Germany, Israel, Poland, the United Kingdom and the U.S.
3. The Hotelling rule requires that the real rate of return to a resource owner equals the social discount rate. When a non-renewable resource is available in known, fixed quantity, the Hotelling rule implies that the net price of the resource should grow at the social discount rate (Perman et al., 2003).

References

Ambec, S. et al. (2016), "Review of international best practices for charges for water management", Toulouse School of Economics, *Background paper for OECD*, 10 October 2016.

Barker, I. (2016), Personal communication.

Margat, J. and J. van der Gun (2013), *Groundwater around the World: A Geographic Synopsis*, CRC Press/ Balkema, Taylor and Francis, London.

OECD (2015a), *Water Resources Allocation: Sharing Risks and Opportunities*, OECD Studies on Water, OECD Publishing, Paris, *http://dx.doi.org/10.1787/9789264229631-en*.

OECD (2015b), *Drying Wells, Rising Stakes: Towards Sustainable Agricultural Groundwater Use*, OECD Studies on Water, OECD Publishing, Paris, *http://dx.doi.org/10.1787/9789264238701-en*.

Olmstead, S. (2010), "The economics of managing scarce water resources", *Review of Environmental Economics and Policy*, Vol. 4(2), pp. 179-198, *http://dx.doi.org/10.1093/reep/req004*.

Perman, R. et al. (2003), *Natural Resource and Environmental Economics*, Pearson, Addison Wesley, 3rd edition.

UK Environment Agency (2016), "Managing water abstraction", UK Environment Agency, Bristol.

Wheeler, S.A. et al. (2016), "Lessons to be learned from groundwater trading in Australia and the United States", *Integrated Groundwater Management*, A. Jakeman (ed.), *http://dx.doi.org/10.1007/978-3-319-23576-9_20*.

PART II

Case studies of groundwater allocation in practice

Groundwater Allocation
Managing Growing Pressures on Quantity and Quality

PART II

Chapter 3

Overview of case studies

This chapter provides an introduction to nine case studies of groundwater allocation in Denmark; Tucson, Arizona; Kumamoto, Japan; Mexico; the Upper Guadiana Basin, Spain; Texas; France; India and North China). It briefly summarises the challenges related to groundwater allocation examined and the elements of the "Health Check" discussed in each case study.

Groundwater allocation poses numerous challenges related to managing both the quantity and quality of the resource, conditioned by the magnitude and type of groundwater use, interactions with surface water bodies and impact on groundwater-dependent ecosystems. The cases presented in the chapter present a range of policy responses put in place to address these challenges in various contexts. The selection of cases was driven by the aim to examine a broad range of groundwater allocation issues, in particular those that were relatively less well-developed in previous work. These include the reallocation of groundwater for environmental purposes and among different types of users; the use of economic instruments, such as abstraction charges and groundwater markets; interactions between quality and quantity aspects of groundwater management; long-term groundwater abstraction limits and the use of proportional pumping restrictions; artificial groundwater recharge; and innovative approaches to the collective management of groundwater allocation.

Table 3.1 provides an overview of the key issues examined in each of the case studies using elements in the *Health Check for Water Resources Allocation* as a framework. The *Health Check* is presented in detail in Chapter 2, along with the policy guidance reflecting good

Table 3.1. **Case studies illustrating the OECD Water Resources Allocation Health Check in practice**

	Denmark	Tucson, Arizona, US	Kumamoto, Japan	Mexico	Upper Guadiana Basin, Spain	Texas, US	France	Gujarat, India	North China
Check 1. Accountability mechanisms in place for the management of allocation		✓				✓	✓		
Check 2. Legal status for all water resources (surface and ground water and alternative sources of supply)					✓	✓			
Check 3. Understanding the availability of groundwater resources and possible depletion	✓	✓	✓	✓					
Check 4. Abstraction limit ("cap") reflecting *in situ* requirements and sustainable use	✓			✓		✓	✓		
Check 5. Approach to enable efficient and fair management of the risk of shortage that ensures water for essential uses				✓					
Check 6. Arrangements in place for dealing with exceptional circumstances (such as drought or severe pollution events)							✓		
Check 7. Process for dealing with new entrants and for increasing or varying existing entitlements					✓		✓		
Check 8. Mechanisms for monitoring and enforcement, with clear and legally robust sanctions	✓					✓			
Check 9. Water infrastructures in place for the allocation regime to function effectively		✓							
Check 10. Policy coherence across sectors that affect allocation	✓		✓	✓		✓		✓	
Check 11. Clear legal definition of water entitlements	✓				✓	✓			
Check 12. Abstraction charges	✓								
Check 13. Obligations related to return flows and discharges									
Check 14. Allowing water users to reallocate water among themselves							✓	✓	✓

practice. The case studies reflect the diversity of contexts and policy responses, demonstrating the importance of tailoring policies to specific conditions. While many of the cases attest to the challenges of groundwater depletion and related negative impacts that remain, the cases nevertheless illustrate the combination of policies that, taken together form an allocation regime, and can, when properly enforced, provide numerous levers to influence the behaviour of groundwater users and ensure the sustainable management of this valuable natural asset.

The case of Denmark provides an example of a comprehensive allocation regime, combining time-bound entitlements, a cap on total abstraction which accounts for environmental needs, economic instruments (volumetric water and wastewater tariffs, taxes, as well as a groundwater abstraction charge) and a well-developed monitoring network. The range of measures in place to protect groundwater quality is of particular importance, as groundwater provides nearly all drinking water in Denmark.

As a rapidly growing desert city that has been heavily reliant on groundwater, Tucson, Arizona in the U.S. provides an example of how developing a diversified water resources portfolio along with water banking and demand management has helped to eliminate groundwater mining as of 2015. Tucson's storage and recovery programme allowed for the water utility to overcome early challenges in integrating new surface water supplies into the system due to quality issues. The case also highlights the importance of flexibility in groundwater allocation and of concerted stakeholder engagement.

The case of Kumamoto, Japan provides an illustration of how a payment for ecosystem services (PES) scheme developed between industrial users and farmers to provide financial incentives for groundwater recharge. The scheme managed to raise groundwater recharge substantially, helping to ensure security of supply for industrial and other groundwater users. Based on this success, the scheme has steadily expanded.

The case studies of both Mexico and Spain examine how concerns about environmental degradation due to groundwater depletion have spurred policy efforts to reallocate water for environmental purposes. In Mexico, groundwater depletion due to uncontrolled pumping has resulted in substantial land subsidence, increased costs of urban and rural water supply and caused the deterioration of groundwater quality. Attempts to exert greater control over pumping have been stymied by weak enforcement. The adoption of the 2012 standard for determining environmental flows was a positive step towards securing water for the environment, however, ambiguity and lack of coherence in national legislation pose challenges to the standard's successful application. In Spain, irrigated agriculture in the Upper Guadiana Basin spurred remarkable socio-economic development, although sharply increased groundwater abstraction resulting in a major decline in the water table. This severe drop negatively impacted several wetlands in the basin, including the famed Tablas de Daimiel National Park, a Ramsar site, which provided valuable ecosystem services (fisheries, crabbing, orchards) to the surrounding population. Over decades, Spanish authorities have put into place policies and legal changes to shift groundwater from private property to a resource managed under the public domain and established pumping quotas. While monitoring and enforcement has been a challenge, these efforts have helped to move from a severely over-abstracted situation towards greater control over abstraction, thereby contributing to the gradual recovery of the aquifer. An ambitious plan to reallocate water to higher value uses and towards environmental purposes has not been fully implemented due to very high costs and budget constraints. However, groundwater levels have recovered the basin, in large part due to high precipitation in recent years, contributing to wetland restoration.

The cases covering examples from Texas, France, India and China highlight how issues related to groundwater allocation for irrigation have been addressed in diverse settings. In some areas of Texas, the Ogallala Aquifer has been subject to depletion for over a half century, resulting in subsidence, brackish intrusion as well as posing a risk to irrigated agriculture and hence, the local economy. Groundwater conservation districts have proved to have a positive impact on the level of groundwater depletion, yet have given rise to conflicts with private property claims, making authorities more reluctant to limit pumping permits in cases where this may result in costly litigation and compensation claims. In the Texas Panhandle, the "50/50"conservation scheme provides a good example of concerted and rigorous long term planning to explicitly account for intertemporal allocation and provide an incentive for farmers to adopt water conservation practices.

In France, the government has instituted a novel institution, the *organismes uniques de gestion collective* (OUGCs), or single collective management bodies, to allow water users to take on the task of allocating a fixed abstraction limit among themselves. Yet, implementation has faced numerous challenges. The OUGCs have sparked strong controversy due to the conflictual relations between those exercising the tasks of the OUGCs and those that are meant to benefit from them (irrigators), as well as decision-making procedures which seem to limit the influence of some stakeholders. Furthermore, farmers have notably reacted to the fact that their individual, permanent water entitlements have been replaced by a collective quota. Also, a lack of clarity regarding key aspects in the legislation, including with regards to sanctioning and the judicial relation between the OUGCs and the farmers, has lead to further lack of support of the collective management model.

In India, where electricity subsidies provide a perverse incentive to pump groundwater, a scheme to ration electricity for the agricultural sector has reduced groundwater use and the cost of electricity subsidies. In North China, severe groundwater depletion presents a threat to the region's food production and economic development. Informal groundwater

Box 3.1. **"Over-exploited": A contested term**

Several of the case studies refer to situations where groundwater resources have been considered "over-exploited". It is important to note that this is a contested term and there is no generally shared interpretation among groundwater specialists. It is employed divergently in different settings, depending on what is considered a normal or acceptable exploitation path.

From an economic perspective, the definition of "over-exploitation" should go beyond simply considering abstraction versus recharge. For example, mining groundwater in non-renewable aquifers to generate capital and invest in the future can be preferable to preserving the stock as such. To some extent, over drafting aquifers may lead to tremendous gains for farmers and communities by later increasing their capacity to adapt to future water constraints (OECD, 2015).

Thus, the definition of "over-exploitation" should be interpreted as a state where the economic, social and environmental costs from a certain level of abstraction exceed the benefits (Garrido and Llamas, 2007). This would imply considering a system in a dynamic cost-benefit analysis, which has merit but also faces challenges. In practice, water management bodies define quantitative reference states to which they compare groundwater levels. Some countries even define multiple water table threshold levels for intervention.

Source: OECD, 2015; Margat and van der Gun, 2013; Garrido and Llamas, 2007.

markets emerged as a response to the privatisation of wells, allowing for increased groundwater access for farmers that lacked the means to install their own wells. The markets are influenced by level of groundwater scarcity, with increased scarcity leading to expanded groundwater market activity. Because electricity tariffs in China are set based on metered consumption, the depth from which groundwater is pumped determines the costs of operating a tube well. When pumping costs are higher, water sellers as well as buyers tend to optimise their groundwater consumption, at least in terms of their private use.

References

Garrido, A. and M.R. Llamas (2007), "Lessons from intensive groundwater use in Spain: Economic and social benefits and conflicts", in Giordano and Villholth (eds.), *The Agricultural Groundwater Revolution: Opportunities and Threats to Development*, CABI, Oxford.

Margat, J. and J. van der Gun (2013), *Groundwater around the World: A Geographic Synopsis*, CRC Press/ Balkema, Taylor and Francis, London.

OECD (2015), *Drying Wells, Rising Stakes: Towards Sustainable Agricultural Groundwater Use*, OECD Studies on Water, OECD Publishing, Paris, *http://dx.doi.org/10.1787/9789264238701-en*.

Groundwater Allocation
Managing Growing Pressures on Quantity and Quality

PART II

Chapter 4

A comprehensive allocation regime in Denmark

This chapter examines groundwater allocation in Denmark, which provides an example of a comprehensive allocation regime, combining time-bound entitlements, a cap on total abstraction which accounts for environmental needs, economic instruments and a well-developed monitoring network. The case study also highlights the importance of measures in place to protect groundwater quality in Denmark, given groundwater's importance as a drinking water source.

Groundwater is the major source of water supply for drinking water, agriculture and industry

Denmark is the only country in the European Union (EU) that uses untreated groundwater for more than 99% of water use, including drinking water (Joergensen and Stockmarr, 2009). In 2014, a total of 735 million m^3 of groundwater was abstracted in Denmark, out of which 425 million m^3 was for non-irrigation purposes (Danmarks Statistik, 2015; Thorling et al., 2015). Agriculture, forestry and fishery consume close to 50% of the total groundwater use in Denmark, 30% is consumed by households and 8% by industry (Danmarks Statistik, 2015). The volume abstracted for irrigation tends to fluctuate significantly from year to year, whereas there has been a slight and steady decrease in the quantities withdrawn for other purposes over the last decade (Danmarks Statistik, 2015; Thorling et al., 2015).

Over the past several decades, groundwater abstraction in Denmark took on increasing importance as an alternative to over-exploited surface water resources[1]. Surface water is generally quite limited due to the country's flat topography. During the 1970s and 80s, Danish surface water resources became over-exploited, as a result of increased household use, the discharge of wastewater to surface water and increasingly dry summers. In response, the government gradually prohibited the direct abstraction of surface water (GEUS, n.d.), with consumption falling substantially over the next decades to about 12 million m^3 consumed in 2014 (Thorling et al., 2015), less than 2% of the amount of groundwater abstracted. Given its importance in the provision of drinking water, protecting the quality of groundwater is also a vital concern (Box 4.1).

Box 4.1. **Protecting groundwater quality in Denmark**

Since the 1970s, there have been concerns related to the quality of Denmark's groundwater resources, which pose a threat to drinking water safety and to the available resource pool (GEUS, 2016). The contamination of the Danish groundwater is due to nitrates from farming, chemicals from old waste dumps and oil tanks, toxic materials from enterprises, and pesticides (GEUS, n.d.). Although both industries and households contribute to nitrate emissions, the rise in nitrate concentrations in groundwater appears to be closely associated with the increasing use of fertilisers (Joergensen and Stockmarr, 2009; GEUS, n.d.). Danish authorities have promoted groundwater quality with a range of measures, including wastewater taxes, improvements in wastewater treatment facilities, taxes and regulations on pesticides and nitrogen fertilisers, targeted protection via municipal action plans for public water supplies as well as an extensive groundwater monitoring network.

Source: GEUS, 2016; GEUS, n,d. ; Joergensen and Stockmarr, 2009.

Groundwater and surface water abstraction is regulated through entitlements granted by municipalities, which must be renewed periodically. The quantity abstracted must be measured and reported to the authorities annually. For groundwater, irrigation entitlements are valid for a maximum of 15 years, whereas entitlements for water utilities can be for a period of up to 30 years (GEUS, n.d.). The Environmental Act states as a general objective that the total volumes of groundwater abstracted should not undermine water-dependent ecosystems' compliance with defined environmental targets (Thorling et al., 2015).

Water pricing as an instrument for demand management and financing

While Denmark has a long tradition of water consumption metering and consumer charges for water supply and waste water treatment (Pedersen, 2016), the EU Water Framework Directive (WFD) encouraged the Danish government to further develop its water pricing system (Hydropolitical Academy, 2014). Full cost recovery for wastewater collection and treatment has been a legal requirement in Denmark since 1992 (OECD, 2012). Full cost recovery also now applies to water supply (NCM, 2006). Water prices in Denmark are particularly high compared to other countries[2].

The average water tariff for Danish consumers has risen steadily over the past two decades, increasing by 350% per m^3 between 1989 and 2012[3] (Hydropolitic Academy, 2014). Figure 4.1 depicts how the annual average household water bill has evolved since 2005 and the breakdown of the bill comprised of water tariffs, wastewater charges and taxes. Denmark is one of the few countries that use a two-part tariff structure, consisting of a flat fee and a charge based on metered consumption[4] (NCM, 2006). The water bill accounts for approximately 1.6% of annual income of Danish families (GEUS, n.d.).

Figure 4.1. **Gradual rise in average Danish annual household water bill, 2005-15**

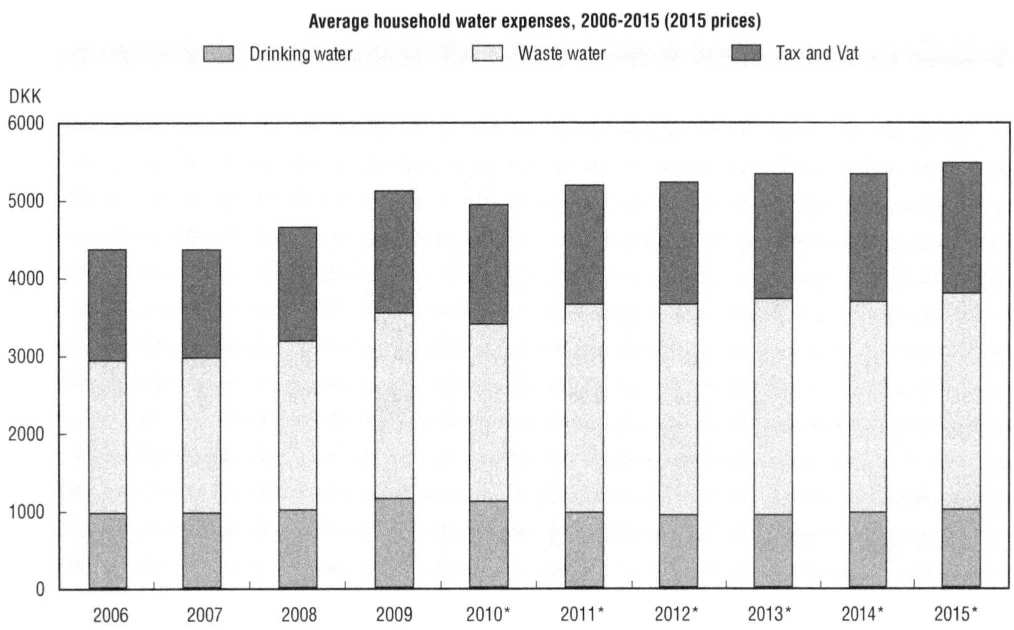

Source: DANVA, 2015. Note: 2014 prices.

From the water bill paid by consumers, approximately 50% accrues to wastewater companies, 30% consists of taxes that accrue to the government and about 20% goes to drinking water utilities (DANVA, 2015). Taxes consist of both levies for water supply and wastewater and value added tax (VAT).[5] The water supply levy was established in 1994, and has been increased in subsequent years. The tax has an environmental purpose and aims to encourage reduced water consumption (NCM, 2006; Pedersen, 2016). The wastewater tax applies to all direct discharges, such as industries and municipal wastewater treatment plants (NCM, 2006). This tax was first introduced in 1997 and the rate was raised by 50% in 2014 (Retsinformation.dk, 2009). Whereas households pay this tax through the water bill, direct dischargers pay it directly to authorities. The charge is proportional to pollution load, and applies to nitrogen, phosphorus and biochemical oxygen demand (NCM, 2006).[6] By providing an incentive for pollution abatement, the wastewater tax contributes to the protection of groundwater quality.

While taxes for water supply and wastewater are applied at a standard rate across country, the water and wastewater tariffs charged by water utilities and wastewater companies vary considerably (NCM, 2006). Households in the areas with the highest tariffs pay six times as much as the ones with the lowest tariffs (Dilling, 2007). This variation is mainly due to structural differences related to the provision of services, including abstraction costs, size and centralisation of customers and maintenance costs, rather than differences in operational efficiency (DANVA, 2015)

Denmark is one of only three European countries (together with France and the United Kingdom) with an abstraction charge for groundwater use (Berbel et al., 2005). The abstraction charge was introduced by the 1994 Green Tax Reform.[7] The charge can be deducted from the farmers' value-added tax proceeds, but is still considered to have an impact on irrigators' groundwater consumption (Berbel et al., 2005).

There is also a tax on water utilities (about EUR 1 per m^3 of water) that aims to reduce water losses via the distribution network, which accounts for about 5% groundwater consumption (Danmarks Statistik, 2015). The tax applies to all water abstracted by utilities, including non-revenue water. Utilities that do not reduce non-revenue water to less than 10% are penalised with additional taxes (NCM, 2006).

Wastewater charges along with the development of more sophisticated wastewater treatment facilities have had a positive impact on groundwater pollution levels.[8] Wastewater charges, which were implemented in 1997, led to a significant decline in the levels of phosphorus (17% annually), nitrogen (5% annually) and organic material (3% annually) in waste water over the first four years (1997-2001) (NCM, 2006). Nevertheless, pollution levels in groundwater have only fallen slightly since then. The increase in the wastewater charge in 2014 aimed to encourage improved treatment of wastewater so as to ensure a decline in the spread of pollutants in the aquatic environment (Retsinformation.dk, 2009). However, it is too early to see the effect of this increase.

Lessons learned

Denmark has developed a comprehensive set of policies for groundwater allocation. The suite of policies to protect groundwater quality is vitally important, considering that Denmark uses untreated groundwater for 99% of its drinking water. Wastewater charges and the improvement of treatment facilities have had an important impact on the aquatic environment, including on the quality and stock of groundwater resources (see Health Check #10, Part I).

The Environmental Act requires that the total volumes of groundwater abstracted should not undermine water-dependent ecosystems' compliance with defined environmental targets (see Health Check #4, Part I). Groundwater and surface water abstraction is regulated through time-bound entitlements granted by municipalities. The duration of entitlements depends on the purpose of use (irrigation or drinking water supply), and on the source of water (groundwater or surface water). Establishing the appropriate duration of entitlements constitute an essential requisite for a clear legal definition of water entitlements (see Health Check #11, Part I).

The total volume of groundwater abstracted must be measured and reported to the authorities annually. This requirement and other measures ensure the continued monitoring of groundwater quantity in Denmark, facilitating the government's, and the public's, understanding of groundwater availability (see Health Checks #3 and #8, Part I).

Furthermore, Denmark's comprehensive system of water tariffs, wastewater charges and taxes provide incentives for reduced pollution and more efficient use of the resources. It also provides the basis for full cost recovery for water supply and wastewater treatment. Groundwater abstraction charges are also in place, including for irrigation (see Health Check #12, Part I).

Notes

1. Certain regions of Denmark experience a much higher pressure on groundwater resources than others. For example, the groundwater in parts of Zealand is largely over-exploited, with the groundwater table in some areas dropping 10-15 meters since predevelopment.
2. In 2007/08, consumer charges for water services in Denmark were higher than in any other OECD country (TASC, 2013).
3. The average price per cubic metre of water, including VAT, is now DKK 63.24 (EUR 8.5) for a typical household (DANVA, 2015).
4. Since the late 1990s, water utilities are legally required to ensure that all properties recently connected to the public water supply are metered, allowing for water supply to be charged based on volumetric rate.
5. The VAT rate of 25% applies to the water and wastewater tariffs in all parts of the country (GEUS, n.d.; NCM, 2006).
6. For treatment plants that receive more than 85% of industrial waste, the wastewater tax applies to volume according to treatment type. Certain high volume consumers are given reduced charges (NCM, 2006).
7. The abstraction charge was fixed at EUR 0.55/m^3.
8. The most significant reduction in groundwater pollution resulted from the development of more efficient and sophisticated wastewater treatment facilities in the ten years following the adoption of The National Plan for the Aquatic Environment in 1987.

References

Berbel, J. et al. (2005), "Water pricing and irrigation: A Review of the European experience", in Molle, F. and J. Berkoff (eds), *Irrigation Water Pricing Policy in Context: Exploring the Gap Between Theory and Practice*, International Water Management Institute.

Danmarks Statistik (2015), "Vandregnskap 2014: Geografi, miljoe og energi" [Water accounts 2014: Geography, environment and energy], Statistics Denmark, *www.dst.dk/Site/Dst/Udgivelser/nyt/GetPdf.aspx?cid=25444* (accessed 16 July 2016).

DANVA (2015), "Water in figures", *http://reader.livedition.dk/danva/171/1* (accessed 16 July 2016).

Dilling, S. (2007), "Enorme forskelle paa vandpriser", [Huge differences in water prices], *http://politiken.dk/forbrugogliv/boligogdesign/energi/ECE441871/enorme-forskelle-paa-vandpriser/*.

GEUS (n.d.), "Water supply in Denmark", *www.geus.dk/program-areas/water/denmark/vandforsyning_artikel.pdf* (Grönwall).

GEUS (2016), "Grundvand og politik", [Groundwater and policy], *www.geus.dk/DK/popular-geology/edu/viden_om/grundvand/Sider/gv06-dk.aspx* (accessed 11 July 2016).

Hydropolitic Academy (2014), "Why water price is so high in Denmark", *www.hidropolitikakademi.org/en/why-water-price-is-so-high-in-denmark.html*, (accessed 11 July 2016).

Joergensen, L.F. and J. Stockmarr (2009), "Groundwater monitoring in Denmark: characteristics, perspectives and comparison with other countries", *Hydrogeology Journal*, June 2009, Vol. 17/4, pp 827-842.

Nordic Council of Ministers (NCM) (2006), "The use of Economic Instruments in Nordic and Baltic environmental policy 2001-05", *TemaNord*, 2006:525, Copenhagen.

OECD (2012), *A Framework for Financing Water Resources Management*, OECD Studies on Water, OECD Publishing, Paris, *http://dx.doi.org/10.1787/9789264179820-en*.

Pedersen, P.G. (2016), Chief advisor, Unit of Water Resources, Agency for Water and Nature Management, Ministry of Environment and Food of Denmark, personal correspondence.

Retsinformation.dk (2009), "2008/1 LF 204", *www.retsinformation.dk/Forms/R0710.aspx?id=124561* (accessed 9 July 2016).

Thorling, L. et al. (2015), "Grundvandsovervaagning 1989-2014", [Groundwater surveillance 1989-2014], GEUS, *www.geus.dk/DK/water-soil/monitoring/groundwater-monitoring/Documents/g-o-2014.pdf*.

Groundwater Allocation
Managing Growing Pressures on Quantity and Quality
© OECD 2017

PART II

Chapter 5

Managing scarce groundwater resources to ensure long-term supply in Tucson, Arizona

This chapter examines groundwater allocation in Tucson, Arizona. Tucson provides an example of how developing a diversified water resources portfolio along with water banking and demand management has helped to eliminate groundwater mining. The case also highlights the importance of flexibility in groundwater allocation and of concerted stakeholder engagement.

Intensive groundwater pumping has led to depletion and land subsidence

In the state of Arizona in the U.S., groundwater provides approximately 43% of the total water supply (Towne and Jones, 2011). In the city of Tucson, as much as 88% of the total water demand was met by groundwater resources as of 2002 (Tucson Water, 2015). The strong dependence on groundwater has resulted in depletion, with groundwater levels declining by 90-150 meters since predevelopment in Tucson and the surrounding area. Consequently, land subsidence of approximately 3.8 meters as compared to 1940 levels has been observed (Ponce, 2006).

Tucson, a desert city, is also subject to a severe risk of surface water shortage. Historical average rainfall has been about 300 millimetres annually. The Colorado River is over-allocated among seven U.S. states and Mexico, and the Colorado River Basin has experienced drought for 14 years. Further, downscaled climate models project that the region will become hotter and possibly drier (Megdal, 2014).

Most residents in and surrounding Tucson are served by Tucson Water, which is a public water utility under the auspices of city authorities (Megdal, 2014). Out of its 709 000 clients, as of 2012, approximately 25% are commercial and industrial water users with the residential users accounting for the remainder (Megdal, 2014).

New surface water sources were introduced to reduce the pressure on groundwater

In response to groundwater depletion, the federal government funded the construction of a 540 km long lined and open canal in Arizona, called the Central Arizona Project (CAP). The CAP was built by the U.S. Bureau of Reclamation and completed in the early 1990s. Operations and repayment are the responsibility of the Central Arizona Water Conservation District, an elected body established by law. Water from the Colorado River is pumped into the CAP from near sea level to a maximum elevation near Tucson of about 730 m. Built to transport approximately 1 850 million m^3 of water annually, the CAP is the largest consumer of electricity in Arizona (Megdal, 2014). In addition to providing a much needed alternative to groundwater, the introduction of CAP water was a way to ensure consistency with the safe-yield management goal for the region. Tucson was granted the largest municipal allocation (approximately 178 million m^3 per year) within the CAP system (Megdal, 2014).

Historically reliant on the region's good quality groundwater, which did not require much treatment prior to delivery to customers, integration of the CAP water through direct delivery required the construction of a large, centralised treatment plant. This was built using a combination of rate-payer charges collected in advance of operation and revenue bond financing. In 1992, Tucson Water delivered treated CAP water to half of its consumers. This first real infusion of surface water into the Tucson Water system turned out to be fraught with difficulties. The CAP water had a different chemistry from that of groundwater and travelled in a different direction through old water mains. The corrosivity of the CAP water was particularly challenging. As a result, the calcium coating on the inside of the water pipes dissolved, allowing rust and soil into the home distribution lines. Many of the galvanised pipes in older parts of the city failed and leaked (Wilson, 2016).

Storage and recovery was implemented as an alternative to direct use of CAP water

The damage caused by the CAP water corrosivity when supplied through the regular network, coupled with the utility's hesitancy to acknowledge the problems, led to lack of confidence and customer activism to restrict the way in which this new water source could be used. In order to respond to consumers' opposition to direct delivery of treated CAP water and to the risk of shortage of surface waters, as well as to comply with the relevant legislation, Tucson Water adopted an indirect approach to utilising CAP water. Rather than treating the water in a large treatment facility and then directly delivering the water to its customers, the utility deployed a Storage and Recovery approach (S&R), in compliance with Arizona State regulations (Megdal, 2014).

Arizona state law has authorised the use of aquifers for water storage and groundwater replenishment and S&R programmes make up an important water management tool in many parts of the state (Megdal, 2014). A system of permits and accounting administered by the Arizona Department of Water Resources (ADWR) governs the construction and use of water storage facilities as well as the recovery of stored water. The permit system allows the ADWR to ensure that the recharge is hydrologically feasible and that no harm is done to water and land resources (Megdal, 2014).

The S&R approach allowed utilising CAP water, first by storing it underground, mostly in large, shallow spreading basins, where it mixes with groundwater in the aquifer, and then by recovering it for distribution (OECD, 2015). Tucson Water is not currently delivering its full allocation of CAP water to its customers, but it is taking delivery of the full allocation. Water over and above that needed to supply current demands is being stored underground for future use. Such storage is extremely important to Tucson Water's ability to withstand Colorado River shortage declarations. Through 2013, Tucson Water invested USD 134 million in the facilities required for its S&R system, with another approximately USD 180 million planned. Annual investment is approximately USD 38.6 million (Megdal, 2014).

The implementation of the S&R programme was facilitated by a number of factors

A number of early decisions by the state and local authorities helped lay the foundation for the implementation of the S&R programme. For example, the state of Arizona had the foresight to establish the Arizona Water Banking Authority (AWBA). In anticipation of a Colorado River shortage declaration. The AWBA has been storing CAP water underground in Tucson Water's storage facilities, since 1997. Consequently, Tucson Water has essentially developed a "drought-proof" system, allowing it to rely on its own storage, as well as that of the AWBA, should there be future curtailment of CAP surface water deliveries. Tucson Water can also increase its use of groundwater if or when needed (Megdal, 2014).

Moreover, in the 1960s and 1970s, Tucson purchased some agricultural lands northwest of the city, with the expectation that the water rights associated with the lands would be used to meet Tucson Water's future demands. What was not envisioned at the time was that these lands would become the site of the large storage facilities that are the backbone of today's S&R system. The ownership of these lands enabled Tucson Water to avoid land acquisition costs when constructing the S&R system (Megdal, 2014).

The implementation of the S&R programme was also aided by Tucson Water's engagement of stakeholders in its planning efforts. In addition to being proactive in its outreach to communities, the water utility embarked on a partnership with a local farming entity at the early stages of implementing the programme. The agricultural partner helped construct some water conveyance infrastructure that was used to deliver CAP water to farm lands and to recharge basins. Tucson Water accrued water storage credits, pursuant to state law and ADWR permitting, for the utilisation of CAP water on the agricultural fields in lieu of groundwater. The state's Groundwater Savings Storage programme, which is incorporated in the statutory framework, is a good example of a mutually beneficial and voluntary partnership (Megdal, 2014).

Local water conservation projects, rainwater harvesting and use of grey water have also contributed to reduce demands on the potable water system and promote conservation (Megdal, 2014). Water conservation has long been a focus of Tucson Water's activities. Water banking is recognised as an important strategy for addressing the long-term needs of the region, and the importance of conservation and wise water use has been a consistent component of Tucson Water's public messaging (Megdal, 2014). Demand management is also promoted through water pricing. Water pricing in Tucson is designed to recover costs of providing water, including extraction, diversion, treatment, delivery, debt service, and administrative charges. The pricing structure for residential customers is based on increasing block tariffs. Commercial customers face higher rates in summer than they do in winter, providing a price signal designed to reduce water consumption during periods of scarcity (Megdal, 2014).

In 2015, Tucson Water reported no mined groundwater use: 84% of the water consumption was supplied by CAP water resources through the S&R approach, 10% was reclaimed water and 6% of total water supply came from remediated (cleaned to a very high standard) groundwater (not considered groundwater use by the regulatory authorities), which is fed into the potable water system (TW & CoT, 2015). This reflects the extent to which the introduction of CAP water as well as the S&R approach has succeeded in altering water consumption patterns in Tucson, resulting in a significant decline in the risk of groundwater depletion.

Tucson Water had to work hard to overcome the loss in confidence that was associated with the failed introduction of CAP water to the Tucson community, and learnt the importance of consulting and communicating with its stakeholders. The water utility has made a particular effort to engage stakeholders in the implementation of its 2013 Recycled Water Master Plan, and regularly informs and engages its governing body – the Tucson Mayor and Council (Megdal, 2014). In addition, Tucson Water's capital investment plan undergoes rigorous review by stakeholders. It is funded through revenue bonds, which through 2005 were submitted to City of Tucson voters for approval. It is worth noting that despite the poor community experience associated with introduction of CAP water, City of Tucson voters approved sizable bond issues to fund replacement of large transmission pipelines and other capital needs. The voters understood that replacing old infrastructure was necessary and supported the higher rates associated with USD 380 million in bonds between 1994 and 2005 (Megdal, 2014).

Lessons learned

Tucson Water serves a desert community, which has, in the past, relied on mining groundwater to support a growing population and economy. Arizona took action as early as 1980 to require use of renewable water supplies to reduce groundwater overdraft. The construction of the CAP enabled the importation of new water supplies to the region. Although strict federal drinking water quality regulations prevail at all times, state groundwater regulations allow utilities flexibility in their approach to utilising renewable surface water supplies. This allowed Tucson Water to adapt from direct delivery of CAP water to an S&R system, avoiding the costs of centralised treatment facilities, and enabling the storage of water for future use (Megdal, 2014). This highlights the importance of flexibility in groundwater allocation and demonstrates Tucson Water's ability to develop a portfolio of alternative sources of supply (see Health Check #3, Part I). The financing and implementation of dedicated infrastructure was crucial for the success of the allocation of CAP water through an S&R approach (see Health Check #9, Part I).

By 2015, there was no mined groundwater use in Tucson. Though cities will not be affected in the short term by the shortage conditions on the Colorado River (due in large part to the establishment of the AWBA) the potential for some curtailment of surface water deliveries has underscored the importance of a diversified water resources portfolio. Demand management and water reuse are significant elements of this approach (Megdal, 2014). Tucson Water's experiences have also demonstrated the importance of effective stakeholder engagement (Megdal, 2014) (see Health Check #1, Part I).

References

Megdal, S. (2014), "Managing water for future cities: Case study of the city of Tucson, Arizona USA", Background Paper prepared for OECD (2015), *Water and Cities: Ensuring Sustainable Futures*, Studies on Water, OECD Publishing, Paris.

OECD (2015), *Water and Cities: Ensuring Sustainable Futures*, OECD Studies onWater, OECD Publishing, Paris, *http://dx.doi.org/10.1787/9789264230149-en*.

Ponce, V. (2006), "Groundwater utilization and sustainability", *http://groundwater.sdsu.edu/* (accessed 22 June 2016).

Towne, D. and J. Jones (2011), "Groundwater quality in Arizona: A 15-year overview of the ADEQ ambient monitoring programme (1995-2009)", Arizona Department of Environmental Quality, *www.azdeq.gov/environ/water/assessment/download/1104ofr.pdf* (accessed 14 August 2016).

Tucson Water and City of Tucson (TW & CoT) (2015), "Water production (1940-2015)", figure prepared for the "Tucson Water Clearwater Renewable Resources Facility 15 Years" event, 25 May 2016.

Tucson Water (2015), "Total water demand", figure prepared for the "Tucson Water Clearwater Renewable Resources Facility 15 Years" event, 25 May 2016.

Wilson, W. (2016), Chief Hydrologist at Tucson Water, personal correspondence.

Groundwater Allocation
Managing Growing Pressures on Quantity and Quality

PART II

Chapter 6

Payments for groundwater recharge to ensure groundwater supply in Kumamoto, Japan

This chapter discusses the case of Kumamoto, Japan, which provides an illustration of a payment for ecosystem services scheme to provide financial incentives for groundwater recharge. The case documents how the scheme managed to raise groundwater levels, improving the security of supply for industrial and other groundwater users.

Groundwater depletion and declining recharge in Kumamoto, Japan

Kumamoto[1] is well known for its groundwater abundance and quality. The region is located in the centre of Kyushu, the southern major island of Japan. Kumamoto City, with a population of 730 000, is also the largest Japanese city that provides 100% of its drinking water from groundwater (UNDESA, 2016). Groundwater is also an essential source of water for agriculture and industry in the region (UNESCO et al., 2015).

Kumamoto City is located in the lower part of an aquifer that is recharged by inflow from the Shirakawa River. The surface layer of the aquifer in Kumamoto has a particularly high permeability and groundwater recharge capacity.[2] Today, one third of groundwater recharge in the region is due to the irrigation of paddy fields with water diverted from the Shirakawa River. Rice paddies in the mid-basin of the river can recharge 5-10 times as much as those in other regions of the country (Nishimiya, 2010). Every kilogramme of rice produced in this area is estimated to raise groundwater levels by approximately 20-30 m^3 (UNDESA, 2016).

Despite the exceptionally high recharge capacity of the aquifer, the groundwater level has declined over recent decades and the recharge capacity of the Shirakawa River is forecasted to decline by 6.2% between 2007 and 2024 (Gundimeda and Wätzold, 2010). This decline is due mainly to a government policy to reduce rice production acreage, which has forced farmers to abandon their paddy fields (UNESCO et al., 2015; UNDESA, 2016). The rice acreage-reduction policy, which was introduced in the 1970s, is a supply-restriction arrangement aimed at supporting prices. Approximately 40% of Japan's paddy fields are subject to acreage-reduction (Kazuhito, 2008).[3] The rice reduction policy has resulted in a significant decrease in agricultural irrigation in Kumamoto with a resulting decline in groundwater recharge (UNDESA, 2016; UNESCO et al., 2015).

The decline in groundwater levels is also a result of increased pumping, as well as rapid urbanisation. An increase in asphalt and concrete surfaces has weakened the ground's capability to absorb water, hence to recharge groundwater (UNDESA, 2016; UNESCO et al., 2015; MoE, 2010). Urbanisation has also led many farmers to abandon rice production, thus further impacting the level of groundwater recharge (ICLEI, 2013).

Promoting groundwater recharge through payments for ecosystem services

To reverse groundwater depletion, a programme for payment for ecosystem services (namely, groundwater recharge) was launched in 2003 as a result of direct negotiations between local farmers and the Kumamoto Technology Centre (Kumamoto TEC), a subsidiary of Sony Semiconductor Kyushu (Hayashi and Nishimiya, 2010). Kumamoto TEC aimed to promote the recharge of the large volumes of groundwater consumed by its semiconductor plant with the objective of becoming "water neutral" (MoE and KC, 2016). The PES scheme allowed the company to pay farmers (per square kilometre per 30 days of flooding) to recharge groundwater by voluntarily flooding fields that had been converted from irrigated rice fields to crop fields. This was be done with water from the Shirakawa River during fallow periods (MoE and KC, 2016; UNESCO et al., 2015; Hayashi and Nishimiya, 2010).

Kumamoto's city government joined the PES programme in 2004 and took a number of complementary actions to encourage groundwater conservation (MoE and KC, 2016). They re-named the PES scheme "Project to Flood" and included it as part of its Water Conservation Plan running from 2004 to 2009 and renewed for the period 2009-13 (Hayashi and Nishimiya, 2010). In parallel, the city government revised the Groundwater Preservation Ordinance, which declared groundwater as a common good that should be conserved (UNESCO et al., 2015). The city also signed a ten year agreement with neighbouring towns, ensuring cross-municipal co-operation for groundwater conservation. The agreement involved the expansion of Project to Flood and the protection of watershed forest in the upper Shirakawa Basin (UNDESA, 2016).

The programme has since expanded, with the Council for Sustainable Water Use in Agriculture[4] (CSWUA) taking on the task of ensuring that flooding is being carried out (UNDESA, 2016; Hayashi and Nishimiya, 2010). Furthermore, several other private companies have committed to fund the recharge programme together with the local authorities. In general, the primary motivation of companies to join the Project to Flood is their interest in preventing groundwater depletion so as to secure sufficient amounts of groundwater for their business activities in the future (MoE and KC, 2016).

The city government and the private companies that have committed to the Project to Flood provide farmers with payment in exchange of their contribution to groundwater recharge. The level of payment is calculated based on an estimate of the preparation and management costs of flooding (Hayashi and Nishimiya, 2010). The CSWUA distributes the payments to farmers (MoE and KC, 2016; UNESCO et al., 2015). In order to receive payments for flooding, farmers have to be located in Kumamoto or in one of the neighbouring municipalities (Ozu-Machi or Kikuyo), use water from the Shirakawa River for irrigation, carry out the flooding for one to three months between May and October before or after cultivation of crops (Kumamoto Water Life, n.d.).

Project to Flood[5] has also inspired related initiatives, including a voluntary scheme to encourage organic rice production. Launched by an agricultural cooperative in Kumamoto in partnership with the city government, the initiative encourages rice cultivation with reduced fertiliser and pesticide use in the fields flooded under Project to Flood. It is supported by local companies, universities and consumers, who buy the "eco-rice" at a slightly higher price than conventional rice (MoE, 2016).

An increase in groundwater recharge

The PES programme in Kumamoto has contributed to increasing groundwater recharge. Figure 6.1. illustrates how groundwater recharge supported by the programme has exceeded water consumption at Kumamoto TEC in most years. In addition to recharging groundwater, the flooding helps limit the negative impact of diseases, weeds and insects (Nishimiya, 2010, UNDESA, 2016).

In 2013, Project to Flood was recognised as the year's best practice in water management by the United Nations Water Decade Programme on Advocacy and Communication and the UN World Water Assessment Programme (UNDESA, 2016; ICLEI, 2013). Kumamoto City's first ten year agreement on the implementation of Project to Flood was completed in 2013, and a second ten year agreement has been initiated. The local government has signalled its interest in continuing the project thereafter (MoE and KC, 2016).

Figure 6.1. **Groundwater recharge has exceeded water consumption at Kumamoto TEC, 2003-15**

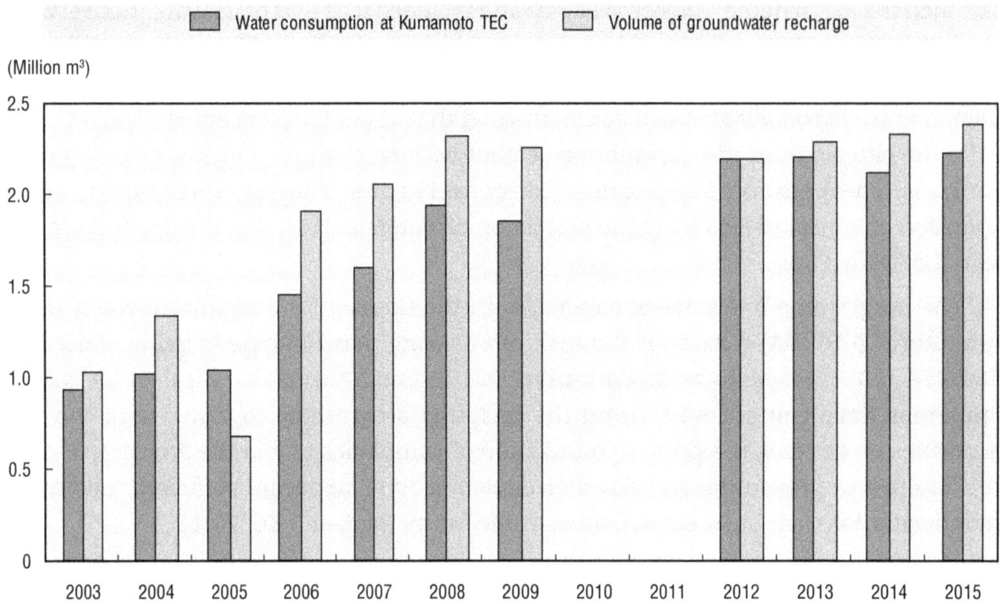

Note: No data available for 2010 and 2011.

Source: Author, based on data from Sony Semiconductor Corporation (2016) *www.sony.net/SonyInfo/csr_report/environment/site/biodiversity/kumamoto.html*. Notes: No data available for 2010 and 2011.

Lessons learned

The case of Kumamoto is one of several examples of cities taking innovative action towards groundwater conservation (OECD, 2015). The case illustrates how a payment for ecosystem services scheme can reverse the groundwater depletion. It also illustrates the importance of policy coherence across agricultural, urban and water policies (see Health Check #10, Part I).

Initially launched by the private sector in partnership with farmers, the scheme later expanded to include local government. The integration of the PES scheme into the local government's broader groundwater management policies has allowed for a more sustained response as well as broader collaboration with an increased number of stakeholders from the public and private sectors as well as civil society. The stakeholders demonstrated a solid understanding of the availability of groundwater resources in the area, the challenges associated with depletion and the possibility to augment supplies through recharge (see Health Check #3, Part I). The programme has facilitated the restoration of groundwater levels and demonstrates how such schemes can provide effective incentives for groundwater recharge while providing greater security of supply for groundwater users.

Notes

1. Kumamoto is both the name of a city and a region.
2. This is due to geographic features of the aquifer as well as the establishment of paddy fields more than 400 years ago (UNDESA, 2016).
3. Approximately 40% of Japan's paddy fields are subject to acreage-reduction (Kazuhito, 2008).
4. The CSWUA consists of local municipalities, land improvement districts and agricultural co-operatives.

5. Project to Flood also involves awareness raising programmes, and since 2008, local governments designate three "Water Saving Months" per year. During these months, daily water consumption per capita is publicly reported and water saving devices are promoted (ICLEI, 2013).

References

Gundimeda, H. and F. Wätzold (2010), "Payments for ecosystem services and conservation banking", In TEEB, *The Economics of Ecosystems and Biodiversity for Local and Regional Policy Makers*, *www.teebweb.org/media/2010/09/TEEB_D2_Local_Policy-Makers_Report-Eng.pdf*.

Hayashi, K. and H. Nishimiya (2010), "Good practices of payments for ecosystem services in Japan", *EcoTopia Science Institute Hayashi Laboratory Policy Brief*, No. 1, Nagoya, Japan, *www.esi.nagoya-u.ac.jp/syupan/ESI%20IGES%202011april%20re%20for%20web.pdf*.

ICLEI (2013), "Kumamoto's 'Project to flood' wins UN-Water Award", *www.iclei.org/details/article/kumamotos-project-to-flood-wins-un-water-award.html* (accessed 7 July 2016).

Kazuhito, Y. (2008), "The pros and cons of Japan's rice acreage-reduction policy", The Tokyo Foundation, *www.tokyofoundation.org/en/articles/2008/the-pros-and-cons-of-japans-rice-acreage-reduction-policy*.

Kumamoto Water Life (n.d.), "Groundwater protection", *www.kumamoto-waterlife.jp/list_html/pub/detail.asp?c_id=25&id=9&mst=0&type* (accessed 18 July 2016).

Ministry of the Environment, Japan (MoE) (2010), "Payment for ecosystem services: Best practices from Japan: Conserving water by recharging groundwater in Kumamoto", *www.biodic.go.jp/biodiversity/shiraberu/policy/pes/en/water/water03.html* (accessed 7 July 2016).

Ministry of Environment in Japan and Kumamoto City authorities (MoE and KC) (2016), personal correspondence.

Nishimiya, N. (2010), "Offsetting industrial groundwater consumption through partnerships between industry and farmers", *The Economics of Ecosystems and Biodiversity*, *www.teebweb.org/wp-content/uploads/CaseStudies/Offsetting%20industrial%20groundwater%20consumption%20through%20partnerships%20between%20industry%20and%20farmers.pdf* (accessed 21 July 2016).

OECD (2015), *Water and Cities: Ensuring Sustainable Futures*, OECD Studies onWater, OECD Publishing, Paris, *http://dx.doi.org/10.1787/9789264230149-en*.

Sony Semiconductor Corporation (2016), "Feature: Working on Groundwater Recharge Projects", *www.sony.net/SonyInfo/csr_report/environment/site/biodiversity/kumamoto.html* (accessed 18 February 2017).

UNDESA (2016), "Basin wide groundwater management using the system of nature: Kumamoto city, Japan", Water for Life, UN-Water Best Practices Award 2013 edition: Winners, *www.un.org/waterforlifedecade/winners2013.shtml* (accessed 2 September 2016).

UNESCO et al. (2015), "Global Diagnostic on Groundwater Governance", UNESCO, the International Hydrological Programme, the World Bank, the Food and Agriculture Organisation, the Global Environment Facility and the International Organisation of Hydrogeologists, *www.groundwatergovernance.org/fileadmin/user_upload/groundwatergovernance/docs/GWG_DIAGNOSTIC.pdf*.

Groundwater Allocation
Managing Growing Pressures on Quantity and Quality

PART II

Chapter 7

Enforcement challenges and efforts to implement environmental flow requirements in Mexico

This chapter discusses groundwater allocation challenges in Mexico. It documents how the government's attempts to exert greater control over groundwater pumping have been stymied by weak enforcement. It also documents efforts to secure water for environmental purposes via standards for environmental flows.

Over-allocation of groundwater resources has led to severe depletion

The allocation of groundwater in Mexico faces significant challenges due to resource scarcity and overdraft, widespread unauthorised use, as well as weak enforcement of national water legislation. On a national level, groundwater supplies 77% of urban water use, 50% of industrial and 33% of agricultural use (Cornett, 2014). In the state of Guanajuato, close to 100% of industrial and domestic water demand is met by groundwater (Foster et al., 2004). Over-allocation of the already scarce groundwater resource pool in Mexico has led to severe depletion. This has dire implications for both rural and urban water supply, as operational and replacement costs have increased significantly. Depletion also results in deterioring groundwater quality (e.g. increased salinity) and land subsidence (amounting to 2-3 cm per year in some areas) with damaging effects on public infrastructure and private property (Foster et al., 2004; Shah, 2014).

Abstraction bans and attempts to regularise users have failed to limit groundwater pumping

In the past, occasional pumping bans were used a key management tool to limit groundwater abstraction. Between 1948 and 1983, eleven pumping bans were issued in Guanajuato, and the National Water Commission (CONAGUA) attempted to enforce three periods of state-wide water well drilling ban during the 1990s. However, the bans were ineffective and the number of wells in Guanajuato has continued to expand rapidly since the 1960s (Foster et al., 2004; Shah, 2014; CONAGUA, 2016).

While groundwater entitlements formerly were tightly linked to land rights, the adoption of the 1992 National Water Act and subsequent amendments to national legislation brought about a reform of groundwater entitlements and regularisation of users. Today, all new and existing water users are legally required to be registered in the Public Register of Water Rights and to be assigned a quantitative water entitlement by CONAGUA. Users are obliged to install meters and report pumped quantities of groundwater to CONAGUA (CONAGUA, 2016; Shah, 2014).

The regularisation of groundwater users was supposed to help enforce drilling permit requirements and bans. However, the requirements regarding metering, reporting and groundwater entitlements are only minimally enforced, and unauthorised groundwater use remains widespread. Poor enforcement and lack of sanctioning is primarily due to lack of local operational resources and failure to mobilise user co-operation (Foster et al., 2004; Shah, 2014). Despite the existence of river basin organisations (RBOs), which are decentralised bodies of CONAGUA, co-ordination of water management across levels of government remains a challenge (OECD, 2013; CONAGUA, 2016). The enforcement of groundwater legislation is also hampered by politicians' challenging trade-off between the control of groundwater abstraction and the desire to attract farmers' votes (Shah, 2014).

Multi-stakeholder platforms have faced numerous challenges

Seeking ways to strengthen water management, the Mexican government established various multi-stakeholder platforms during the 1980s-90s, including river basin councils (RBCs). The RBCs are independent bodies, which are meant to enhance institutional co-ordination as well as the relations between institutions and users of surface and groundwater (Millington, 2006). Twenty years after their creation, RBCs are still not fully operational, and their impact remains contested. In theory, the RBCs are supposed to allow for stakeholders' independent management of water resources in each river basin, but in practice, they are primarily advisory bodies with very little power (Cornett, 2014, OECD, 2013).

The 1992 Water Act also gave impetus to the establishment of technical committees for groundwater (COTAS) in selected aquifers (OECD, 2013). COTAS are non-profit, civil associations created to foster self-regulation of groundwater withdrawal among users (Foster et al., 2004; OECD, 2013). In reality, actual user participation remains minimal and none of the 81 COTAS nation-wide has succeeded in implementing effective ways to reverse groundwater depletion (Molle and Wester, 2009; Cornett, 2014; CONAGUA, 2016). This is due to a lack of human resources, infrastructure and reliable information about well owners. The COTAS do not physically control the extraction infrastructure; thus they can not physically restrict withdrawals by well owners, and have to rely on their goodwill (Molle and Wester, 2009).

A national standard for environmental flows, which still needs to be implemented

Given the limited effects of the new entitlement regime and the multi-stakeholder platforms, Mexico has explored other means to sustain water resources. In 2000, CONAGUA adopted an official national standard requiring the establishment of methods for determination of annual average availability of national waters. NGOs and civil society constituted an important driver for subsequent work on the development of methods for determining environmental flows. Notably, the Alliance of the World Wildlife Fund for Nature and the Gonzalo Río Arronte Foundation reviewed different methodologies for determining environmental flows and explored how to reach an agreement with water users and adopt a legal act in this regard (Barrios-Ordóñez et al., 2015). Inspired by these efforts, the Program of the Environmental and Natural Resources Sector 2007-12 explicitly recommended that the government publish an official standard for the determination of environmental flows, which was done in 2012 (Cornett, 2014) The standard defines environmental flows as the flow rate or minimum volume needed in receiving bodies or the minimum flow of natural discharge of an aquifer in order to protect the environment and the ecological balance of system (OECD, 2015). The official standard is a means to regulate demand and supply for groundwater, and implies a legal recognition of the ecosystem as a legitimate user of water (Rodriguez, 2013).

Two key elements in the standard are the scientific principles of natural flow regime and biological condition gradient (Barrios, n.d., 2012). Natural flow regimes consist of five critical components that regulate ecological processes in river ecosystems. The alteration of these components can result in ecosystems and biological integrity degradation (Poff et al., 1997). The biological condition gradient refers to a conceptual, scientific framework for the interpretation of biological response to increasing effects of stressors on aquatic ecosystems (USEPA, 2016). Based on these two principles, the ultimate objective of the Mexican official standard is to match flow recommendations to available resources and capacity, ecosystem importance and conditions, and the anticipated extent for hydrologic alternation with water resources and infrastructure development (Conservation Gateway, n.d.). The official standard is not only concentrated on site or project level, but encompasses local procedures seeking to apply the Ecological Limits of Hydrologic Alteration (ELOHA) framework (Conservation Gateway, n.d.). The ELOHA framework requires stakeholders and decision-makers explicitly to evaluate "acceptable risk" as a balance between the perceived value of the ecological goals and the economic costs involved, taking into account the scientific uncertainties between ecological responses and flow alteration (OECD, 2015; Poff et al., 2011).

Lessons learned

The enforcement of groundwater legislation remains a challenge for Mexican authorities along with reforming perverse subsidies, such as *Tarifa* 9, a preferential tariff for electricity to pump groundwater for rural users, undermining policy coherence (OECD, 2013) (see Health Check #10, Part I). As a result, groundwater depletion is exacerbated and the resource pool remains over-exploited. The adoption of the 2012 standard for determination of environmental flows has the potential to help adapt groundwater allocation to sustainable levels of groundwater use. By proposing a variety of methodological approaches to environmental flow estimation, the standard allows for the adaptation of methods based on available input and resources (Barrios, n.d.). It also recognises the importance of not fixing a standard minimum environmental flow, but adjusting to different hydrologies across the country (Barrios-Ordóñez et al., 2015).

The standard for determining environmental flows can help to define an abstraction limit (a cap) reflecting sustainable use (see Health Check #4, Part I). It can also be used to identify and protect water reserves. Mexico's National Water Reserves Programme (NWRP), was developed in parallel with the 2012 standard, and seeks to identify potential water reserves where natural flows can be secured. The NWRP has facilitated the examination of 732 basins nation-wide, classifying all watersheds into four categories, based on their water balance and ecological importance. This helps policy-makers obtain a better understanding of the availability of groundwater resources and ensuring environmental uses are secured (see Health Checks #3 and #5, Part I).

The successful implementation of environmental flows as part of the groundwater allocation regime is hampered by a number of obstacles, including elements in the National Water Act. Ambiguity and lack of coherence in national legislation weaken the application of the standard. For example, the legislation gives stakeholders at river basin level the freedom to prioritise water for agriculture and livestock above environmental flows. Legislation also fails to determine whether "environmental use" is one out of many purposes for which water can be granted in an entitlement, or a preliminary restriction of the volume of water available for all entitlements. Moreover, the lack of control of abstraction and the large number of groundwater users complicates the enforcement of environmental flows. As long as the effectiveness of multi-stakeholder platforms for water management remains limited, it will be challenging to reach consensus among users about the importance of restricting groundwater abstraction volumes. Legal reforms may be necessary in order to allow for further harmonisation of the National Water Act and the national standard for environmental flows.

References

Barrios, E. (n.d.), "Environmental flows in Mexico, now a national standard: The recognition of the environment as the only water provider", *Solutions for Water*, *www.solutionsforwater.org/solutions/environmental-flows-in-mexico-now-a-national-standard-the-recognition-of-the-environment-as-the-only-water-provider-3* (accessed 20 May 2016).

Barrios-Ordóñez, J. et al. (2015), "Programa nacional de reservas de agua en México: experiencias de caudal ecológico y la asignación de agua al ambiente" [National program of water reserves in Mexico: experiences of ecological flow and the allocation of water to the environment], Inter-American Development Bank, *https://publications.iadb.org/handle/11319/7316?locale-attribute=es* (accessed 20 May 2016).

Conservation Gateway (n.d.), "ELOHA case study Mexico", *www.conservationgateway.org/ConservationPractices/Freshwater/EnvironmentalFlows/MethodsandTools/ELOHA/Pages/ELOHA-case-study-Mexico.aspx* (accessed 16 July 2016).

CONAGUA (2016), Personal correspondence with Member of the International Cooperation Department, Daniel Rivera, Deputy Manager of Geohydrological Information Systems, Domingo Silva, and Manager of Groundwater, Ruben Chavez Guillen.

Cornett, V. (2014), "Limitations and opportunities for environmental flow implementation under current Mexican law and policy", *Water Law Review*, Vol. 17/10, pp. 223-266.

Foster, S. et al. (2004), "The 'COTAS': Progress with stakeholder participation in groundwater management in Guanajuato, Mexico", *Groundwater Management: Lessons from Practice, Case Profile Collection*, Number 10, GW Mate, The World Bank, *http://siteresources.worldbank.org/INTWRD/Resources/GWMATE_English_CP_10.pdf* (accessed 12 August 2016).

Millington, P. (2006), "Integrated river basin management: from concepts to good practice: case study 5: the Lerma-Chapala River Basin, Mexico", World Bank Institute, *www-wds.worldbank.org/external/default/WDSContentServer/WDSP/IB/2007/10/19/000310607_20071019123528/Rendered/PDF/411690MX0Lerma1ase1study1501PUBLIC1.pdf* (August 4 2016).

Molle, F. and P. Wester (eds.) (2009), *River Basin Trajectories: Societies, Environments and Development*, CABI and IWMI, MPG Book Groups, Bodmin.

OECD (2015), "Water resources allocation in Mexico", country profile, *www.oecd.org/mexico/Water-Resources-Allocation-Mexico.pdf* (accessed 13 July 2016).

OECD (2013), *Making Water Reform Happen in Mexico*, OECD Studies on Water, OECD Publishing, Paris, *http://dx.doi.org/10.1787/9789264187894-en.*

Poff, L. et al. (2011), "Ecological limits of hydrologic alteration: Environmental flows for regional water management", *www.conservationgateway.org/Documents/brochure-english(1).pdf* (accessed 25 June 2016).

Poff, L. et al. (1997), "The natural flow regime", *BioScience*, Vol. 47/11, pp. 769-784, *http://wec.ufl.edu/floridarivers/RiverClass/Papers/natflow_paradigm.pdf.*

Rodriguez, S. (2013), "Environmental flows Mexican standard: implications on hydropower", WWF Mexico, *http://inogo.stanford.edu/sites/default/files/WWF%20-%20Environmental%20Flows%20Mexican%20Standard%20-%20Implications%20of%20Hydropower_0.pdf* (accessed 6 August 2016).

Shah, T. (2014), "Groundwater governance and irrigated agriculture", *TEC Background Papers, No. 19*, Global Water Partnership Technical Committee, Global Water Partnership, *http://www.gwp.org/globalassets/global/toolbox/publications/background-papers/gwp_tec_19_web.pdf*, (accessed 18 July 2016).

United States Environmental Protection Agency (USEPA) (2016), "A practitioner's guide to the biological condition gradient: a framework to describe incremental change in aquatic ecosystems", EPA-842-R-16-001, USEPA, Washington, DC, *www.epa.gov/sites/production/files/2016-02/documents/bcg-practioners-guide-report.pdf.*

Groundwater Allocation
Managing Growing Pressures on Quantity and Quality

PART II

Chapter 8

Enforcement and budget challenges for groundwater reallocation in the Upper Guadiana Basin, Spain

This chapter summarises efforts by Spanish authorities to slow groundwater depletion in the Upper Guadiana Basin. The case study discusses the policies and legal changes put into place to shift groundwater from private property to a resource managed under the public domain as well as efforts to reallocate water to higher value uses, including the environment.

Groundwater depletion spurred socio-economic development, with negative environmental impacts

The Upper Guadiana Basin[1] in Spain has been subject to intensive groundwater use for agriculture irrigation for several decades. The intensive groundwater withdrawal in the basin was due to a great extent to a regional policy originating in the 1970s providing local farmers with subsidies to pump groundwater to irrigate drylands in the area. Irrigated agriculture spurred remarkable socio-economic development in the region, although the policy led to a fourfold increase in abstraction during the 1970s and 80s, largely exceeding the recharge of the groundwater resource. This resulted in a drop in the water table of between 20 and 30 meters, and as much as 50 meters in some places with negative implications for the natural environment (Rodríguez-Cabellos, 2016; López-Gunn et al., 2011).

Because of the proximity and connection of groundwater and surface water resources in the Upper Guadiana Basin, the drop in groundwater levels negatively affected several wetlands in the basin. This included the Tablas de Daimiel National Park,[2] which used to be regarded as a landmark and was included in the national Ramsar list (Rodríguez-Cabellos, 2014). Its natural springs and wetlands provided the surrounding population with an ecosystem that allowed it to sustain itself in terms of fisheries, crabbing and orchards. During the 1980s, the flooded surface area of Tablas de Daimiel saw a decline from 6 000 Ha to less than 1 000 Ha (Global Water Partnership, n.d.). Only 20% of the original wetland areas in the national park remained in 2010 (López-Gunn et al., 2011).

Early efforts to stem groundwater depletion

With the introduction of the 1985 Water Act, groundwater was brought under the public domain and a system of entitlements with abstraction limits was introduced (Box 8.1). In the Upper Guadiana Basin, the number of groundwater users who registered their wells under the grandfathering provision of the Water Act by far exceeded the available renewable resources (MoA, 2016).[3] This contributed to the observed continued decline of the groundwater table.

Due to the sharp decline in the water table,[5] the Guadiana River Basin Management Agency declared the Western Mancha aquifer as provisionally "over-exploited" in 1987, and made this decision definitive in 1994 (Sanchez-Carillo, 2010). The agency introduced an exploitation regime involving several restrictions for groundwater use, such as a prohibition of drilling of new wells and deepening of existing ones, compulsory formation of Water User Associations, and further reducing groundwater abstraction quotas per hectare for entitlement holders.[6] The restrictions provoked strong opposition (Rodríguez-Cabellos, 2016; Rodríguez-Cabellos, 2014; Global Water Partnership, n.d.).

However, the River Basin Management Agency lacked capacity and financial means to ensure compliance with the new rules. Thus, over-abstraction of existing water entitlements continued, along with an increasing number of cases of illegal drilling (Rodríguez-Cabellos, 2016). A survey carried out among 70% of irrigators in the basin in 2005 showed that actual

Box 8.1. **From private ownership to public domain: Key elements of Spain's 1985 Water Act**

The groundwater allocation regime in Spain has evolved over time from a system of private groundwater ownership to one of increasing control by the State to limit overall abstraction and manage groundwater entitlements. For over a century (from 1879 to 1985), Spanish law defined groundwater ownership as private property and recognised the right of land owners to abstract unlimited amounts of groundwater; in other words, the rule of capture applied (MoA, 2016; Rodríguez-Cabellos, 2014). In 1985, the Water Act was adopted, which declared groundwater to be in the public domain and introduced a system of groundwater entitlements. Entitlements are valid for a maximum of 75 years, mainly granted to individuals and are temporarily transferrable in special drought situations (OECD, 2015a). In cases where an aquifer has been declared as "over-exploited" (see discussion of the term "over-exploited" in Chapter 1, Box 1.1) or at risk of not achieving good status, all users are required to organise themselves into groundwater user associations (Shah, 2014; MoA, 2016). There are no abstraction charges for groundwater in Spain (OECD, 2015b).

The 1985 act had a grandfathering provision, which allowed for groundwater users who already possessed and operated wells at the time the law was enacted to continue abstracting water under the same conditions (e.g. diameter and depth of the well, pumps conditions, etc.) as before as long as they registered their wells. When registering their wells, users were assigned a maximum abstractable volume, which varied by region and also by irrigated crop type.[4] Entitlement holders could then decide whether they wanted 1) to continue with no temporal limit, keeping exploitation conditions unchanged, or 2) in the case where they wanted to change some of the exploitation conditions, they could transform their entitlement into a time-bound entitlement valid until 2035 when it would be turned into a time-bound entitlement (*concesión*) aligned with the 1985 law. (MoA, 2016; Rodríguez-Cabellos, 2016).

The 1985 Water Act also allowed for authorities to declare depleted aquifers as over-exploited (Rodríguez-Cabellos, 2014). It addition, the act renewed the status of the River Basin Management Agencies (*Confederaciones Hidrográficas*), first introduced in 1927, which are in charge of water resource planning and development, issuing groundwater entitlements and monitoring water quality and quantity (Rodríguez-Cabellos, 2016).

Source: MoA, 2016; OECD, 2015a; OECD, 2015b; Rodríguez-Cabellos, 2014; Shah, 2014.

groundwater abstraction that year amounted to 54 million m^3 above the quantity authorised by the agency, which was 170 million m^3 (De Stefano and López-Gunn, 2012).

In addition to the declaration of over-exploitation, the Autonomous Government in Castilla-La Mancha introduced a ten-year Income Compensation Plan in 1992 to reduce abstraction and contribute to wetland recovery, while compensating farmers for income losses. In exchange for economic compensation, farmers were required to use less water (or no water at all), abandon water-intensive crops in favour of water-efficient crops, and reduce fertiliser and pesticide use (Rodríguez-Cabellos, 2014). In addition, a plan to restructure vineyards encouraged a shift from groundwater use for herbaceous crops (e.g. maize and beet) to less water-intensive crops, such as vineyards (Rodríguez-Cabellos, 2014).

Promoting ecological restoration: The Special Plan for the Upper Guadiana Basin

Although the policies introduced in the 1980s and 90s managed to reduce abstraction to some extent, severe overdraft still persisted. To address this, the Upper Guadiana Special Plan (UGSP) was approved by the Spanish Council of Ministers in 2008. The plan set out to obtain social and ecological restoration primarily through reallocating water entitlements on equity and efficiency grounds. To do so, the plan sought to reduce abstraction to 200 million m^3 per year by 2027 and to raise groundwater levels in order for the Tablas de Daimiel to once again become the natural discharge and overflow from the aquifer (López-Gunn et al., 2011). A Water Rights Exchange Centre was established in order to purchase groundwater entitlements and reallocate them for regularisation and environmental purposes (Garrido and Llamas, 2009). Thirty percent of purchased entitlements were to be reallocated to less water intensive and more economically viable agricultural production. In practical terms, this meant that groundwater entitlements would be purchased from cereal farmers and reallocated to farmers who were illegally using groundwater to irrigate fields with vines, olives, vegetables and horticultural products[7] (López-Gunn et al., 2012). The production of these crops yields more labour and added value per drop of water, and is less exposed to competition from other countries (López-Gunn et al., 2012; López-Gunn et al., 2011). The remaining 70% of purchased entitlements were intended for environmental restoration (López-Gunn, 2012).

The UGSP also involved pumping restrictions and closing down of illegal drilling, implemented with the aid of satellite remote sensing and flow metering devices (López-Gunn et al., 2012; Garrido and Llamas, 2009). Moreover, the plan sought to convert grandfathered water entitlements into entitlements regulated by the 1985 Water Law, and reduce quotas to be consistent with available water resources. The plan also introduced a comprehensive environmental programme, including (re-)forestation measures, in addition to measures for strengthened management, awareness-raising, improved water supply and sanitation, as well as social and economic development (Martinez-Santos et al., 2014; Garrido and Llamas, 2009; Rodríguez-Cabellos, 2014).

The total budget for the UGSP amounts to EUR 5.2 billion for the period 2008-27 (OECD, 2015c; PEAG, n.d.) (Figure 8.1). The total budget for the purchase of water entitlements (EUR 810 million) was set based on the price of purchase of groundwater entitlements, which was fixed to a maximum of EUR 10 000 per hectare, serving as a compensation for the farmers (López-Gunn et al., 2011; 2012).[8] A 2011 assessment of the cost-effectiveness of the UGSP confirms that the average price of water entitlements remained approximately EUR 5/m^3 (López-Gunn et al., 2012).

The plan has been criticised for being overly ambitious and costly, as well as for its failure to comply with the principle of full cost recovery for water services set by the EU WFD (Martinez-Santos et al., 2014). The UGSP was supposed to be financed mainly by the Spanish central government, but funding has fallen short due in part to the sharp economic contraction due to the global financial crisis. Consequently, the purchase of groundwater entitlements, notably for environmental restoration, has not materialised to the level that was envisaged (Martinez-Santos et al., 2014). In fact, as much as 81% of the entitlements that had been purchased by 2011 were reallocated to illegal irrigators, primarily vine farmers (López-Gunn et al., 2012). As a result, the UGSP has not had the expected impact in terms of environmental outcomes. However, the redistribution of entitlements to unauthorised users has also had positive implications for the restoration of water resources, leading to a reduction in abstraction of up to 17.03 hm^3, as those who sold their entitlements would cease their water consumption (Rodríguez-Cabellos, 2016).

Some of the shortcomings of the plan were addressed by measures in the Guadiana District Management Plan (GDMP) for the period 2009-15. Among the measures proposed

Figure 8.1. **Budget of the Upper Guardiana Special Plan, 2008-27**

(EUR millions)

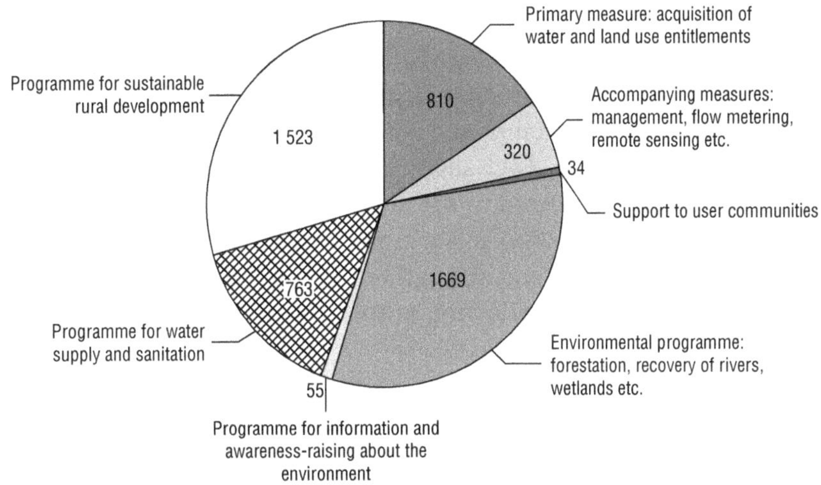

Source: Authors, adapted from PEAG, n.d.

was an entitlements exchange system that would allow for private contracts of entitlement transfers based on the new legislation, facilitating a sort of water trade. However, this has not been implemented (Rodríguez-Cabellos, 2016). Further, the GDMP proposed implementing a new system for declaration of risk in groundwater bodies in the Upper Guadiana basin, for use in cases where there is a chance that good status of water quality and quantity, in accordance with the EU WFD, will not be achieved. In cases where a shortage risk declaration has been made, restrictions to abstractions without compensation can be implemented. This provides a strong measure for restricting use (MoA, 2016; Rodríguez-Cabellos, 2014).

Studies show that the UGSP has succeeded in strengthening the regulatory environment in the basin, and that monitoring and sanctioning are being carried out to a much larger extent than before (López-Gunn et al., 2012). As a result of the UGSP, monitoring based on satellite remote sensing of groundwater bodies is now in force; however, the installation of water-metering devices has not begun (Rodríguez-Cabellos, 2016). According to the government, the number of existing groundwater entitlements is still three times the available resource, the latter being determined based on a definition of sustainable status of groundwater resources and related ecosystems (MoA, 2016).

The combination of policy measures adopted to improve the groundwater allocation regime in the Upper Guadiana Basin has driven a shift towards higher value crops, and triggered an important reduction in the total volume of abstraction. Overall abstraction of groundwater in La Mancha has been reduced from 640 km^3 annually in the mid-1980s to 240 km^3 per year currently (Rodríguez-Cabellos, 2014). The current total abstraction volume is reported to be compatible with the available resources as defined by the GDMP to be necessary to achieve good status by 2027 (Rodríguez-Cabellos, 2016). Furthermore, from 2009 to 2012, a 21 metres increase in groundwater levels was observed. In 2011, the Tablas de Daimiel National Park increased its flooded area from 0 to 2000 ha; a faster and larger recovery than ever previously observed and the water level is now close to 1979 records (Rodríguez-Cabellos, 2014). However, the rise in water table is primarily due to unusually large quantities of precipitation in the period 2006-10 (Martinez-Santos et al., 2014).

Lessons learned

Despite the rise in groundwater levels resulting from policy measures as well as fortuitous precipitation, the purchase of water entitlements under the UGSP has not lived up to the ambition to reallocate a majority of the entitlements to environmental protection purposes. The Upper Guadiana Basin case illustrates that the buy-out of entitlements has the potential to ameliorate environmental conditions if the entitlements are retired or allocated to environmental uses, but usually at a high cost. The case demonstrates the opportunities and challenges related to the altering of existing water entitlements (see Health Check #7, Part I). For the UGSP to succeed in meeting its 2027 objectives, the plan should continue to be reviewed and strengthened, possibly with extended use of risk declarations and new measures allowing authorities to restrict abstraction without compensation.

The transition from the former legislation to the new Water Act is not yet completed, but has nevertheless allowed for a shift from private ownership of groundwater resources towards management under the public domain (see Health Check #2, Part I) and the development of a clear legal definition of water entitlements (see Health Check #11, Part I). Groundwater abstraction control will be gradually strengthened up until 2035, as the grandfathered entitlements from the former water law are converted into regular entitlements, aligned with the allocation regime dictated by the 1985 law.

Notes

1. The basin measures approximately 16 700 km^2, or 2% of Spanish territory. It is located in the south-west of the Iberian Penisula (Rodríguez-Cabellos, 2016; López-Gunn et al., 2011).
2. The Park was established in 1973.
3. Today almost 95% of water entitlements in Upper Guadiana are grandfathered rights, half of which have no temporal limit and half of which will expire in 2035 (MoA, 2016).
4. The abstractable volume was set to different levels in different regions; in the Upper Guadiana Basin, it was set to 4 278 m^3/ha for herbaceous crops, and 2000 m^3/ha for vineyards.
5. In the Ojos del Guadiana springs, which is the Western Mancha Aquifer discharge point, the water table dropped substantially during the period 1979-1993, and fell further during the 1990-95 drought. In these years, severe environmental damage occurred (Rodríguez-Cabellos, 2014).
6. Entitlement quotas for herbaceous crops were limited to 2 000 m^3/ha, and for vineyards, quotas were limited to 1 500 m^3/ha.
7. Recipients of reallocated entitlements also had to meet a set of other criteria, including being under 40 years old, farming as a main occupation, and currently using groundwater for irrigation without formal rights. These criteria reflected the social aspects of the UGSP, and were intended to promote equity by redistributing access to water among farmers (López-Gunn et al., 2012; López-Gunn et al., 2011).
8. The value of groundwater entitlements to be purchased was determined by an economic study carried out by a consortium for the development of the UGSP. In determining the upper bound of the purchase price, the study considered the overall water availability in the basin and considered the relationship between water use and the gross added value of crops (MoA, 2016). Then, several Public Offers for Acquisition of water entitlements were organised, during which entitlement holders offered to sell their entitlements at an offer price (no greater than the upper bound set by the study). The best offers, which fulfilled the criteria for acquisition were selected and constituted the final price of sale (MoA, 2016).

References

De Stefano, L. and E. López-Gunn (2012), "Unauthorized groundwater use: Institutional, social and ethical considerations", *Water Policy*, Vol. 14, pp. 147–160.

Garrido, A. and M.R. Llamas (eds.) (2009), *Water Policy in Spain*, DRD Press, Taylor & Francis Group, Boca Raton.

Garrido, A. and M.R. Llamas (2007), "Lessons from intensive groundwater use in Spain: Economic and social benefits and conflicts", in Giordano and Villholth (eds.), *The Agricultural Groundwater Revolution: Opportunities and Threats to Development*, CABI, Oxford.

Global Water Partnership (n.d.) "Spain: managing water demand in the Upper Guadiana Basin, Case #18", *www.gwp.org/Global/ToolBox/Case%20Studies/Europe/Spain.%20Managing%20water%20demand%20in%20the%20upper%20Guadiana%20basin%20(%2318).pdf*.

López-Gunn, E. et al. (2012). "Tablas de Daimiel National Park and groundwater conflicts", In De Stefano, L. and Llamas, M.R., *Water, Agriculture and the Environment in Spain: can we square the circle?*, CRC Press/Balkema, Taylor & Francis Group, London, UK.

López-Gunn, E. et al. (2011), "The impossible dream? The Upper Guadiana System: Aligning challenges in ecological systems with changes in social systems", *Selections from the 2010 World Water Week in Stockholm*, *www.rac.es/ficheros/doc/00841.pdf* (accessed 4 August 2016).

Martinez-Santos, P. et al. (eds.) (2014), *Integrated Water Resources Management in the 21st Century: Revisiting the Paradigm*, CRC Press/Balkema, Leiden.

Ministry of Agriculture in Spain (MoA) (2016), personal correspondence.

OECD (2015a), "Policies to manage agricultural groundwater use: Spain", country profile, *www.oecd.org/tad/sustainable-agriculture/groundwater-country-note-SPA-2015%20final.pdf* (accessed 16 July 2016).

OECD (2015b), "Water resources allocation in Spain", Country profile, *www.oecd.org/spain/Water-Resources-Allocation-Spain.pdf* (accessed 16 July 2016).

OECD (2015c), *Drying Wells, Rising Stakes: Towards Sustainable Agricultural Groundwater Use*, OECD Studies on Water, OECD Publishing, Paris, *http://dx.doi.org/10.1787/9789264238701-en*.

Plan Especial del Alto Guadiana (PEAG) (n.d.), "Special Plan for the Upper Guadiana Basin: Synthesis document", *www.chguadiana.es/corps/chguadiana/data/resources/file/PEAG/0_DOC_SINTESIS.pdf* (accessed 11 July 2016).

Rodríguez-Cabellos, J.A. (2016), Head of Planning Office, Guadiana River Authority, personal correspondence.

Rodríguez-Cabellos, J.Á. (2014), "Towards IWRM in the Upper Guadiana Basin, Spain", in Martinez-Santo, P., M.R. Llamas and M. Aldava (eds.), *Integrated Water Resources Management in the 21st Century: Revisiting the Paradigm*, CRC Press/Balkema, Taylor & Francis Group, London.

Sanchez-Carillo, S. and D. Angeler (eds.) (2010), *Ecology of Threatened Semi-arid Wetlands: Long-term Research in Las Tables de Daimiel*, Springer, Dordrecht, Heidelberg, London and New York.

Shah, T. (2014), "Groundwater Governance and Irrigated Agriculture", *TEC Background Papers*, No. 19, Global Water Partnership Technical Committee, Global Water Partnership, *www.gwp.org/Global/Toolbox/Publications/Background%20papers/GWP_TEC_19_web.pdf*.

Groundwater Allocation
Managing Growing Pressures on Quantity and Quality

PART II

Chapter 9

Long term abstraction limits to conserve groundwater in Texas

This chapter examines how groundwater conservation districts in Texas have had a positive impact on the level of groundwater depletion. However, efforts by authorities to limit groundwater pumping have given rise to conflicts with private property claims in some cases. The case also discusses the "50/50" conservation scheme in the Texas Panhandle, which provides a good example of concerted and rigorous long term planning to explicitly account for intertemporal allocation and provide an incentive for farmers to adopt water conservation practices.

The depletion of the Ogallala Aquifer in Texas has severe implications for the state economy. The Ogallala Aquifer, the largest freshwater aquifer in the U.S., has been subject to depletion in some areas of Texas for over a half century. Since the 1950s, the Ogallala Aquifer in Texas has been pumped approximately six times the estimated rate of recharge to the aquifer (Mace, 2016). The consequent decline in groundwater levels and saturated thickness constitutes a severe threat to the sustainability of irrigated agriculture in Texas, and hence to the local economy (Mace, 2016). Sixty percent of Texas' water supply comes from groundwater, and 40% of the state's total supply is withdrawn from the Ogallala Aquifer (Foster, 2009). The agricultural sector in the Southern High Plains region of Texas fully depends on water for irrigation from the Ogallala Aquifer. The state-wide economic value directly derived from irrigated agriculture in Texas was USD 4.7 billion in 2007 (TWRI, 2012). The economic impact of converting all irrigated acres in the Texas High Plains to non-irrigated dryland farming would constitute an annual net loss of USD 1.6 billion of gross output, USD 616 million of value added and close to 7 300 jobs (TWRI, 2012). Furthermore, groundwater depletion in Texas has resulted in subsidence and brackish intrusion (Foster, 2009).

Groundwater conservation districts as a means to control pumping

In the US, groundwater quality is managed by federal agencies, while the states are responsible for groundwater quantity policies (Foster, 2009). The Texas Supreme Court established the rule of capture as a state-wide principle for groundwater allocation in 1904, allowing landowners to pump unlimited amounts of groundwater underlying their own property for beneficial use (Johnson et al., 2009). The principle has gradually been modified over the years. In 1949, a legislative session authorised the implementation of groundwater conservation districts (GCDs) (Lesikar et al., n.d.). The GCDs have the responsibility, and the right, to protect, preserve and conserve groundwater resources through necessary regulation (Weinheimer et al., 2012). As of 2016, there are 98 confirmed GCDs state-wide. Sixty-one of the GCDs cover single counties, and 37 cover more than one county. A total of 177 out of 254 counties are either fully or partially within a GCD (TWDB, n.d.a).

In 1997, Texas legislature recognised GCDs as the preferred institution for groundwater management, and gave them the authority to manage groundwater through restrictive rules, and to develop and adapt management plans. A 2002 bill authorised the introduction of new policy tools to reduce groundwater withdrawal, and established 16 groundwater management areas for planning and co-ordination of management plans across GCDs (TWDB, n.d.b; Mace et al., 2006). Since 2005, the GCDs have been obliged to develop and present quantified desired future conditions (DFCs) of relevant aquifers to the Texas Water Development Board (TWDB) (Johnson et al., 2009). The DFCs are based on local decisions made by GCDs, typically aided by technical studies. TWDB incorporates the DFCs into a groundwater model to develop a withdrawal amount that the districts can include as a consideration for permitting purposes (TWDB, 2016).

Every five years, the GCDs submit groundwater management plans to TWDB for approval. The plans have to consider relevant management goals, and establish performance standards and measures allowing for the goals to be reached. Further, the plans must include proposed rules and estimates regarding the available groundwater in the district based on the DFCs, annual groundwater abstraction volumes, annual recharge from precipitation, annual discharge from groundwater to springs and surface water, annual exchange of groundwater between aquifers in the district and within each aquifer between districts, projected surface water supply, and projected total water supply and demand according to the most recent state water plan (TWDB, 2016). Key regulatory tools used by the GCDs include well permitting, spacing and tract size requirements, restrictions on out-of-district transfers and withdrawal limitations. Some districts rely on information and behavioural change campaigns rather than regulatory tools. Abstraction charges are also allowed by the Texan legislation, but as of 2009, these have not been used by any of the Texas High Plain GCDs (Foster, 2009; Johnson and Ellis, 2013).

GCDs have a positive impact on depletion, but can give rise to conflicts

Studies show that the GCDs have an overall positive impact on the levels of groundwater depletion in Texas. The groundwater users in GCDs are obliged to consider trade-offs between the present and the future and evidence suggests that considering the temporal allocation of groundwater resources has had a positive overall impact by slowing the rate of groundwater depletion. Many of the areas that are not covered by GCDs are subject to open access problems and experience increased groundwater depletion as a result (Foster, 2009). TWDB reports that on a state-level, most groundwater users generally work within the framework of the management plans and rules of the GCDs. However, in parts of the state, there are tensions between protecting private property rights and the legislative mandate for districts to preserve the resource though pumping limits and well spacing requirements. There is historical and ongoing litigation on these issues, and most observers agree that court cases on the subject will continue (TWDB, 2016). An often cited case is that of *Edwards Aquifer Authority (EAA) vs. Day*, which came to an end in 2012 (Box 9.1).

Box 9.1. **Legal battle over groundwater in Texas: The Day Case**

Mr. Day and Mr. McDaniel (jointly referred to as "Day") owned a piece of land within the Edwards Aquifer Authority (EAA)'s jurisdiction, beneath which a groundwater source flowed under artesian pressure. The previous owner of Day's land had abstracted groundwater for irrigation, both directly from the well and from an impoundment on a creek within the property to which the artesian flow had been directed by a ditch constructed by the landowner. The Edwards Aquifer Authority Act assured landowners who had used groundwater historically for irrigation purposes a minimum permit amount of two acre-feet of production per year per acre irrigated. Thus, on the basis of the historical use of the previous landowner, Day requested a permit to irrigate 700 acre-feet of land with water from the well and the impoundment. The EAA granted Day a permit for 14 acre-feet of groundwater for irrigation withdrawn directly from the well, but denied the request for a larger permit amount, claiming that the water abstracted from the impoundment was surface water, thus owned by the state, and did not constitute historical use of groundwater from the Edwards Aquifer (John and Ellis, 2013; Kulander, 2015).

Box 9.1. **Legal battle over groundwater in Texas: The Day Case** *(cont.)*

Day claimed that the denial of the permit request represented a constitutional taking of property. Thus, he appealed to the state District Court, alleging error by the EAA and seeking damages for condemnation of his groundwater rights. In response, the EAA sued the state, insisting that the state should be liable in the event that the Court found that there was a taking. The District Court ruled for the EAA and granted a take-nothing summary judgment on all of Day's constitutional claims. However, the Court of Appeals reversed the summary judgment and ordered a remand for further proceedings. Subsequently, the case was taken to Supreme Court, which confirmed the ruling of the Court of Appeals, and recognised landowners' property interest in groundwater in place beneath their land, similar to a landowners' vested property right to oil and gas. The court also established that landowners have the right to be compensated for their interest in groundwater, enabling the plaintiffs to proceed on their takings claim (John and Ellis, 2013; Kulander, 2015; Wilder, 2013).

The Day case gave impetus to a number of other takings claims where landowners have required compensation from GCDs, based on the impact of pumping regulations on their investment-backed expectations. Critics are worried that the court's ruling in the Day case will have negative implications for groundwater conservation in Texas, as the GCDs now have to take into account the economic impact on landowners when defining DFCs, in order to avoid costly compensation demands (Wilder, 2013). For example, this is likely to impact the GCDs' position with regard to the use of groundwater for oil and gas operations, such as hydraulic fracturing ("fracking"). Until 2011, hydrocarbon exploration and drilling activities, including fracking, were exempted from permit requirements for groundwater use in Texas. However, the severe 2011-12 drought led a number of GCDs to seek to regulate or prevent the use of groundwater for oil and gas operations. The Day case nonetheless made them reluctant to do so, as they became aware that denial of permits is likely to lead landowners to file litigation seeking compensation (Kulander, 2015; Johnson and Ellis, 2013).

Source: John and Ellis, 2013; Kulander, 2015; Wilder, 2013.

A long-term, flexible approach to limit groundwater abstraction

The GCD in the Panhandle, located in the north of Texas, provides a compelling example of how GCDs can succeed in lowering the volumes of groundwater abstraction. Nearly all (95%) of the groundwater withdrawn in Panhandle is used for irrigation (Johnson et al., 2009). In 1998, the Panhandle GCD introduced a "50/50" management policy, which is a water pumping quota scheme (Weinheimer et al., 2012). The policy was based on a DFC of ensuring that at least 50% of the initial water supply, and saturated thickness, would still be available 50 years later. Limitations for withdrawal were set accordingly. Panhandle GCD chose to start out setting the annual quota to 1.25% of the initial saturated thickness, and to recalculate the quota every five years, based on the evolution of the level of depletion in the aquifer, which is measured in specific wells. The intention was that conservation of groundwater through the 50/50 scheme would allow for a gradual transition from irrigated to dryland cropland in Panhandle (Johnson et al., 2009).

In order to monitor compliance with the 50/50 scheme, the Panhandle GCD adopted procedures to identify study areas where groundwater declines are believed to exceed the annual decline rate set to be consistent with the 50/50 management goal. In the designated study areas, water level data and groundwater production data are evaluated. This evaluation may lead to the establishment of conservation areas where additional metering of groundwater production and possible production limits will be enforced. Despite district rules allowing for the use of financial penalties in case of violation of production limits, the Panhandle GCD has relied on irrigators' voluntary compliance to date.

The groundwater quota scheme has divergent impacts across the district

Evidence shows that the 50/50 policy in the Panhandle GCD has had largely divergent impacts across the district with regards to adoption of conservation practices. The quotas imposed by the scheme, which are even across the district, represent a much bolder ambition for farmers in areas with low initial saturated thickness and high water withdrawal patterns. Hence, these farmers have been obliged to change their farming practices to a larger extent than those with a low withdrawal pattern and initial high saturated thickness underlying their farm (Johnson et al., 2009). In fact, in certain areas in the Panhandle GCD the saturated thickness was projected to be as high as 80% of the initial level after 50 years, meaning that the ambition to conserve 50% of it imposed no actual restriction on farmers in these areas (Weinheimer et al., 2012). In order to be effective even in areas with high initial saturated thickness and to keep all farmers equally responsible for conservation efforts, the policy would have to be adapted to the spatial variations in water withdrawal patterns and heterogeneity in the aquifer (Johnson et al., 2009; Weinheimer et al., 2012).

The divergence in saturated thickness and withdrawal patterns also impact the extent to which farm production and income are affected by the 50/50 policy. Overall farm production in the Panhandle has not been substantially affected by the policy, neither has the overall economy of the district. Nevertheless, farm production has been slightly reduced in areas where the saturated thickness drawdown levels were initially particularly low. Likewise, the farmers in areas with low saturated thickness and high withdrawals are to some extent negatively affected. However, these farmers are very few, and their economic viability has only been slightly altered (Weinheimer et al., 2012).

The adjustable quota scheme offers several advantages

Instead of adjusting the quotas under the 50/50 scheme every five years, the Panhandle GCD could have opted for a model where the annual quota is fixed to 1% of the initial saturated thickness. This would guarantee that water withdrawal would not exceed 50% during the 50 years planning horizon. However, the adjustable quota scheme employed in Panhandle offers several advantages. For example, it allowed for more water use in the early years (1.25%), as compared to a fixed quota scheme (1%). It may seem counter-intuitive to opt for a model that accepts a comparatively higher annual decline in saturated thickness; however, the adjustable quota scheme is likely to turn out more beneficial over time, as it provides incentives for conservation early on and allows for enhanced flexibility over the long term. The adjustable scheme rewards conservation efforts over time by revisiting the quotas every five years. As farmers are aware that quotas are adjusted periodically, they have an incentive to adjust their practices to conserve water in an earlier period, so that this water will be available to them later. Conversely, farmers under a fixed quota scheme know that the quota will remain constant over the fifty years of the policy scheme, no matter if they adopt conservation behaviour or not. Thus, the farmer has no clear incentive to reduce abstraction levels (Johnson et al., 2009). Moreover, the flexibility integrated in the Panhandle scheme allows for readjustment of quotas according to the development of demand and of the aquifer, in cases where this differs from original assumptions (Mittelstet et al., 2011).

A simulated comparison study of the 50/50 policy and a 50 years abstraction charge scheme concluded that the 50/50 scheme is more successful in terms of conserving a larger quantity of water, hence maintaining a bigger part of the saturated thickness. Further, it leads to a greater decline in the irrigated area, thus making it a more effective groundwater conservation tool. Nevertheless, it comes with higher costs for the regional economy, as the reduction in irrigated area leads to a decline in economic activity for input purchases and reduced levels of gross revenues from the agricultural sector. In contrast, since legislation allows for abstraction charges to be set quite low, a charge would have minimal impact on demand and function primarily as a revenue raising instrument for the GCDs. For farmers, an abstraction charge would represents an additional (small) cost applying to every unit of abstracted water, reducing the net income of the farmer. However, as farmers would maintin the size of their irrigated area and level of production, this would not negatively affect the regional economy. If an abstraction charge were set at a higher level, the impact of this instrument could change, depending on the price-elasticity of water demand (Johnson et al., 2009).

Lessons learned

The gradual strengthening of GCDs' responsibility to conserve groundwater resources in Texas has proved to have a positive impact on levels of groundwater depletion. Through the development of DFCs and pumping permits, the GCDs provide long-term exploitation strategies (see Health Check #11, Part I). The GCDs provide important mechanisms for allocation of groundwater resources and are accountable for the overall groundwater management (see Health Check #1, Part I). Nonetheless, the GCDs have given rise to conflicts between private property rights and groundwater conservation, making GCDs more reluctant to limit pumping permits in cases where this may result in costly litigation. The Day case illustrates how a complex legal context can pose challenges for groundwater management (Health Check #2, Part I).

The example of the Panhandle GCD allows for a comparison of different approaches for temporal allocation of scarce groundwater resources. Compared to alternative approaches, the 50/50 scheme in Panhandle appears to be a highly flexible conservation tool, providing an incentive for farmers to adopt water conservation practices over time. The scheme constitutes an effective short- and long-term abstraction limit (a cap) (see Health Check #4, Part I). The scheme also entails clear systems for monitoring of compliance with the conservation policy, as well as provides sanctioning systems, although only used to a limited extent (see Health Check #8, Part I). However, the magnitude of the impact of the scheme on farmers varies considerably across within the district, due to heterogeneity in aquifer characteristics and consumption patterns. The economic impact of the policy reflects the same trend; certain farmers are hit harder than others. As a consequence of the relative success of the 50/50 scheme in Panhandle, similar policies have been adopted in several other GCDs of the Southern Ogallala Aquifer.

References

Foster, J. (2009), "Do Texas groundwater districts matter?", *Water Policy*, Vol. 11, pp. 379-399, *http://dx.doi.org/10.2166/wp.2009.015*.

Johnson, J. et al. (2009), "Water conservation policy alternatives for the Ogallala Aquifer in Texas", *Water Policy*, Vol. 11 (5), pp. 537-552, *http://dx.doi.org/10.2166/wp.2009.202*.

Johnson, R. and G. Ellis (2013), "A new day? Two interpretations of the Texas Supreme Court's ruling in Edwards Aquifer Authority v. Day and McDaniel", *Texas Water Journal*, vol. 4/1, pp. 35-54.

Kulander, C. (2015), "Edwards Aquifer Authority V. Day and Bragg – Predictions on their effects for regulatory takings claims for groundwater used in oil and gas operations", *Baylor Law Review*, vol. 66/3, pp. 472-528.

Lesikar, B. et al. (n.d.), "Questions about Groundwater Conservation Districts in Texas", *Texas Cooperative Extension, http://twri.tamu.edu/reports/2002/2002-036/2002-036_questions-dist.pdf* (accessed 3 August 2016).

Mace, R. (2016), "So secret, occult, and concealed: An overview of groundwater management in Texas", Conference paper, *https://www.twdb.texas.gov/groundwater/docs/2016_Mace_OverviewGroundwater Management.pdf* (accessed 19 June 2017).

Mace. R. et al. (2006), "A streetcar named desired future conditions: The new groundwater availability for Texas", State Bar of Texas. 7th Annual The Changing Face of Water Rights in Texas, May 18-19, *www.twdb.texas.gov/groundwater/docs/Streetcar.pdf*, (accessed 19 June 2017).

Mittelstet, A., M. Smolen, G. Fox and D. Adams (2011), "Comparison of aquifer sustainability under groundwater administrations in Oklahoma and Texas", *Journal of the American Water Resources Association*, 47(2), pp. 424-431, *http://dx.doi.org/10.1111/j.1752-1688.2011.00524.x.*

Texas Water development Board (TWDB) (2016), Personal correspondance with Media Relations Specialist, Kimberly Leggett, Director of Groundwater, Larry French, and Manager of Groundwater Technical Assistance, Rima Petrossian.

Texas Water Development Board (TWDB) (n.d.a.), "Groundwater conservation district facts", *www.twdb.texas.gov/groundwater/conservation_districts/facts.asp* (accessed 4 August 2016).

Texas Water Development Board (TWDB) (n.d.b), "Groundwater Management Areas", *www.twdb.texas.gov/groundwater/management_areas/* (accessed 4 August 2016).

Texas Water Resources Institute (TWRI) (2012), "Status and trends of irrigated agriculture in Texas", *http://twri.tamu.edu/docs/education/2012/em115.pdf* (accessed 4 August 2016).

Weinheimer, J. et al. (2012), "Economic impact of groundwater management standards in the Panhandle Groundwater Management District of Texas: Final report", *www.twdb.texas.gov/publications/reports/contracted_reports/doc/0903580958_Panhandle.pdf* (accessed 12 August 2016).

Wilder, F. (2013), "Come and take it: Court ruling dares regulators to limit pumping", Observer, *www.texasobserver.org/texas-court-upholds-takings-claim-landmark-water-case/*

Groundwater Allocation
Managing Growing Pressures on Quantity and Quality

PART II

Chapter 10

The collective management approach for irrigation in France

This chapter reviews the case of groundwater allocation for irrigation in France, where the government has instituted collective management bodies to allow water users to take on the task of allocating a fixed abstraction limit among themselves. The case documents the key features of the approach as well as numerous implementation challenges.

Collective management bodies as an attempt to reduce over-exploitation of groundwater

France is ranked as the 11th largest user of absolute volumes of groundwater among OECD countries. As of 2013, 63% percent of abstracted groundwater is used for urban purposes, 20% for agriculture and 17% for industry (OECD, 2015). The ownership of groundwater is linked to property ownership; however, the use of groundwater is regulated by the government, and entitlements are required for use. The nature of groundwater entitlements depends on whether or not a basin is classified as a *zone de repartition d'eau* (ZRE) (BRGM, 2016). Introduced in 1994, ZREs are zones where the state has the mandate to exercise a stricter allocation of water resources, due to the structural deficiency of water supply as compared to demand.

Years after the introduction of ZREs, the water-balance remained over-exploited many places across the country. In response to this pressure and in order to meet aims under the EU Water Framework Directive (WFD), the 2006 Water Law, *La Loi sur l'Eau et les Milieux Aquatiques* (LEMA) introduced a new collective model for the allocation of surface and groundwater resources for irrigation purposes: *les organismes uniques de gestion collective* (OUGCs, or collective management bodies) (Loubier and Polge, 2016; Di Franco, 2016). The implementation of these collective management bodies is mandatory in basins that are classified as ZREs, albeit strongly recommended also in other basins. Most of the current collective management bodies were implemented in 2012 or later (Figureau et al., 2012).

An OUGC can best be described as a function or a task, rather than a body. This function can be carried out by a number of different groups or institutions, including agricultural chambers, groups of local irrigators, owners of land used for irrigation, local legal groups or territorial associations. Those wishing to obtain the OUGC mandate apply to the Prefecture, which appoints the most suitable group in collaboration with the local Water Agency and agricultural chamber. The majority of existing OUGCs are run by agricultural chambers, while a few are operated by irrigators' unions (BRGM, 2016; Di Franco, 2016; Figureau et al., 2012). The body appointed as OUGC is initially given a time-bound mandate (three to five years), with the possibility of extension for an unlimited period of time (Garin, 2016).

The OUGCs are in charge of collecting water withdrawal requests from irrigating farmers in a defined water apportionment zone (e.g. a basin), and, based on these requests, propose annual plans for the allocation of the total abstractable volume of water among the irrigators. The Prefecture determines the total abstractable volume of water for the local agricultural sector based on a nationally-defined minimum water flow (BRGM, 2016; Patrice et al., 2013; Loubier and Polge, 2016; Rinaudo and Hérivaux, 2014). In addition, the OUGCs develop multi-annual plans projecting the distribution of water for irrigation over a period of up to 15 years. Both plans are provided to the Prefecture, which, with or without making amendments, approves the plans. It is important to note that the mission of the OUGCs is only to prepare the decisions of the Prefecture, which remains the ultimate authority with regards to allocation of water (Loubier and Polge, 2016).

Before the creation of the OUGCs, groundwater in ZREs was allocated through a system of permanent, individual abstraction quotas. Irrigators could either request abstraction volumes directly from the Prefecture, or place their request with the agricultural chamber, which subsequently would forward a collective demand to the Prefecture. Outside of the ZREs, there were no abstraction quotas, and entitlements were permanent (Di Franco, 2016; BRGM, 2016). The OUGCs now have monopoly with regards to receiving and managing abstraction requests. This implies that farmers have an obligation to inform the OUGC about their abstraction needs, within a date defined by the OUGC (Loubier and Polge, 2016).

Modalities of representation may hinder irrigating farmers' influence

Representation and decision-making procedures within the OUGCs depend on, and are inherited from, the structures of the body exercising the OUGC task. This implies that the length of time during which each member of the OUGC remains in his or her position depends on the electoral characteristics of the structure assuming the role of the OUGC. For example, agricultural chambers organise elections for all representatives every six years, leading to a replacement of the people in the OUGC at the same interval (MEEM, 2016; Di Franco, 2016). Further, this means that there is no guarantee that the local irrigators are represented in the OUGC. An agricultural chamber may represent farmers of diverse kinds – not only irrigating farmers - and may manage a territory that is overlapping with, rather than identical to, that of the OUGC. As a result, it is possible that irrigating farmers from the area managed by the OUGC are not be represented in the general assembly, which makes the decisions regarding allocation criteria and repartition plans for the OUGC.

In practice, however, most OUGCs have established consultative committees that include irrigators. The custom is that these committees propose a set of allocation criteria as well as plans for repartition of water to the general assembly, which generally respect these propositions when making decisions (Loubier and Polge, 2016; Di Franco, 2016; Garin, 2016; Rinaudo and Hérivaux, 2014). Nevertheless, critics argue that the absence of a standard for representation and decision-making procedures in OUGCs may lead to internal and external conflicts over how decisions on water allocation are taken as well as potentially weaken channels of influence for irrigating farmers and the internal legitimacy of the OUGCs (Patrice et al., 2013; Di Franco, 2016; Rinaudo and Hérivaux, 2014).

Several aspects of the allocation regime provoke debate

Once established, the OUGCs develop a set of internal rules that govern the internal functioning of the OUGC as well as the relations between the OUGC and the irrigating farmers in the water apportionment zone. These rules dictate the representation within the OUGC, its internal organisation (e.g. sub-committees, the modalities of consultative committees), the type of information irrigators have to provide to the OUGC and when, rules and criteria for water allocation, procedures for management of conflicts, reactions to rule violations, the internal budget and rules for how to fix a potential fee to be paid by irrigators (Loubier and Polge, 2016; MEEM, 2016; Rinaudo and Hérivaux, 2014). An inter-organisational working group, including members from the French Permanent Assembly of Agricultural Chambers and three national irrigators' unions, has developed a guiding document for OUGCs seeking to establish their internal rules. The document is not prescriptive, but provides best practices as well as raises key issues that should be taken into consideration (Rinaudo and Hérivaux, 2014).

When a new OUGC is established, the local Water Agencies, which are financed by user charges on water, subsidise up to 50% of its staff costs, with a progressive decline over the first five years. After that, the OUGCs are expected to be financially independent with a balanced budget, separated from the general budget of the institution assuming the role of the OUGC. In accordance with Decree 2012-84, the OUGCs are entitled to implement irrigator charges in order to generate revenues (Garin, 2016; Di Franco, 2016; Rinaudo and Hérivaux, 2014). The OUGCs are free to design the charges as they see fit. For example, the charge may consist of a fixed and a variable fee, with the variable portion determined by a number of different parameters. Critics point out that the choice of these parameters can create a number of challenging trade-offs and conflicts (Rinaudo and Hérivaux, 2014). Furthermore, observers emphasise that the legislation does not clearly define the judicial relation between farmers and the OUGCs; hence, raising questions about the basis for imposing charges (Loubier and Polge, 2016; Figureau et al., 2012). Surveys show that OUGCs have struggled to enforce irrigators' fees. Consequently, several OUGCs have found it difficult to cover their costs.

Each OUGC is free to define its own allocation criteria. This can be a highly politicised task, as well as a source of conflict, as it defines which farmers and what farming practices will be prioritised over others. The aspects and principles on which the allocation criteria can be based include historical water use, types of crops, cultivation techniques, soil quality, the economic viability or the environmental impact of cultivation activities, the size of the irrigated surface, and the age of the farmer. For example, whether young, or new, farmers, should be given priority over old ones, and whether those fields needing the most water should be prioritised, or those farmers using water the most efficiently, are some of the many challenging questions that OUGCs must address when determining the allocation criteria. When defining the criteria, all OUGCs have to respect the principle of equity between users. This implies that similar users are entitled to similar – *equal* - treatment. Nonetheless, the OUGCs can easily prevent this principle from becoming an obstacle to their own allocation criteria. For example, by defining users into different categories, based on their own criteria, the OUGCs can justify that irrigators, who initially could appear to be alike, are treated differently with regards to water allocation (Loubier and Polge, 2016; Figureau et al., 2012; Rinaudo and Hérivaux, 2014).

The Prefecture has the right to additionally restrict the total abstractable volume due to droughts or other conditions putting increased pressure on water resources[1]. The OUGCs are in charge of proposing a modified version of the allocation criteria, also called a set of rules for crisis management, allowing for the modification of annual repartition plans when this occurs (Loubier and Polge, 2016; Rinaudo and Hérivaux, 2014). However, critics state that the legislation remains unclear with regards to how the allocation criteria should be altered in times of restriction, as compared to under normal circumstances (Patrice et al., 2013).

The OUGCs are not mandated to monitor farmers' compliance with the repartition plans; this role is assumed by the Prefecture, the *Office National de l'Eau et des Milieux Aquatiques* along with the *Directions départmentales des Territoires* (Loubier and Polge, 2016; Di Franco, 2016; Rinaudo and Hérivaux, 2014; Figureau et al., 2012). However, the legislation states that the OUGCs are entitled to take account of non-compliance. This creates ambiguity with regards to the juridical grounds on which the OUGCs can gather information about compliance, create acceptance for its allocation principles and potentially impose sanctions (Patrice et al., 2013). Questions also remain as to whether potential sanctions should apply in cases where single farmers have exceeded their authorised volume, or only when the total abstractable volume has been exceeded (Loubier and Polge, 2016).

Despite the lack of a clear mandate, various OUGCs have envisaged ways of sanctioning non-compliant farmers, including through reducing their abstraction volumes (Polge and Loubier, 2016). Some of these sanctioning mechanisms have been strongly contested, both because of their potential implications for cultivation and because it remains debatable whether the OUGCs should interfere in cases of non-compliance at all. Many farmers, who for a number of reasons are opposed to the allocation plans developed by the OUGCs in the first place (notably the irrigators' unions) are particularly at odds with the idea of having the OUGC imposing sanctions. The conflictual relationship between irrigators' unions and agricultural chambers in many parts of France, due to divergent interests, has a long history. For example, agricultural chambers may face criticism by the irrigators' unions when they carry out public administrative tasks (MEEM, 2016). Consequently, the members of the OUGCs often find it challenging to be in charge of enforcing sanctions (BRGM, 2016). OUGCs were established with the aim that allocation criteria would be developed in a manner considered legitimate by most farmers and thus, result in the emergence of a self-managed system, reducing the need for sanctions and monitoring. However, the conflictual relations between many OUGCs and the irrigators has prevented this from materialising (Patrice et al., 2013).

Lessons learned

As the implementation of OUGCs is a very recent development, and yet has to take place in some areas, it would be premature to attempt a comprehensive assessment of their functioning. However, some key observations can be made at this early stage. The French government has made substantial efforts to strengthen the conservation and allocation of national water resources through the introduction of OUGCs. The OUGCs were developed with the intention of becoming accountability mechanisms for the management of groundwater allocation at local scale, often basin scale (see Health Check #1, Part I). In principle, the introduction of the OUGCs allows for the adjustment and enforcement of the abstraction limits defined by the prefectures (see Health Check #4, Part I). Furthermore, it has enabled the development of allocation criteria specific to crisis management, ensuring that there are adequate arrangements in place for dealing with exceptional circumstances (see Health Check #6, Part I). In principle, the OUGCs also allow for water users to reallocate water among themselves as a means to improve the allocative efficiency of the regime (see Health Check #14, Part I), as well as to make decisions regarding processes for dealing with new entrants, increasing or varying entitlements (see Health Check #7, Part I).

Nevertheless, in practice, their implementation has, so far, sparked strong controversy due to the conflictual relations between those exercising the tasks of the OUGCs and those that are meant to benefit from them, as well as decision-making procedures which seem to limit the influence of some stakeholders. Furthermore, farmers have notably reacted to the fact that their individual, permanent water entitlements have been replaced by a collective quota. Many irrigators perceive this as an expropriation, and assert that the OUGCs alter historical allocation criteria (Rinaudo and Hérivaux, 2014). The lack of clarity regarding key aspects in the legislation, including with regards to sanctioning and the judicial relation between the OUGCs and the farmers, leads to further lack of support of the collective management model. Several farmers view the OUGCs, which oblige them to respect rules defined by the OUGC, potentially including paying fees, as an arrangement offering few advantages (Loubier and Polge, 2016; Patrice et al., 2013).

In order to ease farmers' resistance to the OUGCs, the latter could benefit from strengthening their communication, so as to explain to the irrigators why the total abstractable volume has been restricted. Many irrigators claim that they do not understand the rationale behind the decline in abstraction volumes (Di Franco, 2016). According to the Ministry of Environment, farmers are reluctant to recognise the scarce nature of water resources, thus resist reductions in abstraction quotas, as well as the OUGCs (MEEM, 2016; Di Franco, 2016). The success of French policies for conservation of water for irrigation depends on a change in farmers' views; a key objective should be to convince farmers that a restriction of aggregate abstraction volumes eventually will offer advantages by reducing the risk of water shortage in the long-term (Rinaudo and Hérivaux, 2014).

While the introduction of OUGCs reflects the government's willingness to delegate decision-making power to farmers, critics argue that the OUGCs have not been endowed with any actual power, as the government still is in charge of appointing the members of the management bodies, determining the total abstractable volume of water, and approving the annual repartition plans (Figureau et al., 2014; Patrice et al., 2013). The effectiveness of the OUGCs could be potentially enhanced by developing a stronger legal framework for the OUGCs, notably with regards to representation and decision-making procedures. Nevertheless, it is too early to draw final conclusions on the functioning and impact of the OUGCs given that the process of implementation still is ongoing.

Notes

1. If the intervention to restrict the total abstractable volume occurs more frequently than during 2 years out of a 10 year period, it is necessary to revise the overall authorisation of the OUGC as it would have been proven to have erred in determining the overall authorisation.

References

Bureau de recherches géologiques et minières (BRGM) (2016), personal correspondence with Coordinator of Research Program on Water & Environmental Economics, Jean-Daniel Rinaudo.

Di Franco, F. (2016), The French Permanent Assembly of Agricultural Chambers, personal correspondence.

Figureau, A.G. et al. (2012), "Gestion quantitative de l'eau d'irrigation en France : Bilan de l'application de la loi sur l'eau et les milieux aquatiques de 2006", [Quantitative management of irrigation water in France: Assessment of the application of the Water and Aquatic Environment Act 2006], *http://infoterre.brgm.fr/rapports/RP-61626-FR.pdf* (accessed 12 August 2016).

Garin, P. (2016), Member of the Research Collective Gestion de l'Eau, Acteurs, Usages, personal correspondence.

Loubier, M. and S. Polge (2016), "Étude sur les règlements intérieurs des Organismes Uniques de Gestion Collective et sur les critères d'allocation de la ressource en eau", [Study on the rules of procedure of the Single Collective Management Organisations and on the criteria for the allocation of water resources], ONEMA and IRSTEA, 2013-15.

Ministry of Environment, Energy and the Sea in France (MEEM) (2016), personal correspondence with Head of the Mission on Water and Aquatic Environments, Jérémy Devaux and colleagues.

OECD (2015), "Policies to manage agricultural groundwater use: France", Country profile, *www.oecd.org/tad/sustainable-agriculture/groundwater-country-note-FRA-2015%20final.pdf* (accessed 16 July 2016).

Patrice, G., L. Sébastien and C. Myriam (2013), "Irrigation individuelle – irrigation collective : état des lieux et contraintes", [Individual irrigation – collective irrigation : state of play and constraints], *Sciences Eaux & Territoires*, Vol. 2(11), pp. 86-89, *www.cairn.info/revue-sciences-eaux-et-territoires-2013-2-page-86.htm* (accessed 11 July 2016).

Rinaudo, J.D. and C. Hérivaux (2014), "Quels instruments pour une gestion collective des prélèvements individuels en eau pour l'irrigation ? ", [Which instruments for collective management of individual water withdrawals for irrigation ?], synthesis of the ONEMA workshop in Paris, 10 February 2014.

Groundwater Allocation
Managing Growing Pressures on Quantity and Quality

PART II

Chapter 11

Co-managing electricity and groundwater allocation in Gujarat, India

This chapter examines efforts to address groundwater depletion in Gujarat, India. This case explores how a scheme to ration electricity for the agricultural sector has reduced groundwater use as well as the cost of electricity subsidies.

Groundwater scarcity and pollution an increasing challenge in India

Groundwater accounts for approximately 60% of agricultural irrigation and 80% of drinking water in India (Cullet, 2014). Close to 60% of Indian states face challenges in terms of groundwater scarcity and/or pollution (Cullet, 2014). In 2004, 28% of India's blocks (nationally recognised administrative units), showed dangerously high levels of groundwater use, as compared to 4% in 1995 (Planning Commission, 2011). Conversely, there are parts of India that suffer from groundwater excess, also posing challenges to allocation and management (Planning Commission, 2011).

While the Government of India has claimed the right to surface water since the 19th century, groundwater has been controlled by private land owners, making it difficult to regulate and protect (Cullet, 2014; Water Governance Facility, 2013). As the Indian law does not recognise the ownership of groundwater, land owners have never had legal private ownership of groundwater resources, but their unlimited right to abstraction has often been interpreted as *de facto* ownership (Water Governance Facility, 2013). However, a number of legislative changes have provided state governments with increasing control of groundwater resources. The federal government's scope to influence groundwater management remains limited, as water is considered a state matter. This was affirmed by the Groundwater Model Bill adopted in 1970, which recognised groundwater as a matter of "local concern". The model bill gave state governments the right to intervene in the management of the resource in order to protect it, albeit to a limited extent (Kaushik, 2016; Cullet, 2014; Planning Commission, 2011). The model bill was updated in 1996 and 2005 with the latest amendments allowing for the regulation of groundwater development[1] in areas notified by the State Ground Water Authority. In 2011, the Model Bill for Conservation, Protection and Regulation of Groundwater was developed, which seeks to include groundwater under the public trust doctrine (Kaushik, 2016). Indian states are encouraged to adopt the 2011 model bill in such a way that it suits the specific conditions and needs of each state, as well as existing institutional and legal framework on a state level (Kaushik, 2016).[2] Only a few states have implemented the new model bill (Cullet, 2014).

Groundwater allocation challenges and policy responses: The example of Gujarat

The state of Gujarat, located on the western coast of India, has historically faced considerable challenges in terms of groundwater allocation. Gujarat has a population of 60 million, out of which 57% live in rural areas (Census India, 2011). As of 2008, approximately 45% of the population depended on agriculture for their livelihood (Shah et al., 2008). More than 77% of water for irrigation in Gujarat comes from groundwater resources and due to worsening scarcity of surface water resources, the pressure on groundwater has increased over the last decades. Except for some of its southern districts, Gujarat is one of the most water-stressed states of India (Bala, 2015).

The state authorities of Gujarat have made significant efforts to respond to local groundwater challenges. It has initiated the enactment of the 2005 Model Bill for

Groundwater, but this has yet to be fully implemented (Kaushik, 2016). It is also one of very few states that have included groundwater regulation in irrigation legislation. The 2013 Gujarat Irrigation and Drainage Act requires that farmers pay a fee to irrigate land with groundwater within a distance of 200 meters from a canal. The legislation also makes it mandatory for farmers apply for a license from the local canal officers in order to construct a tube well or bore well when the depth exceeds 45 meters (Cullet, 2014; Bala, 2015; Desai, 2013). The enforcement of the 2013 act is contested by farmers across Gujarat, who assert that sinking bore wells is the only option that they have in order to meet their water requirements (Desai, 2013). Moreover, enforcement is challenged by the sheer costs of monitoring over one million wells scattered over the state[3] (Parekh, 2014). Another major challenge frequently faced by groundwater authorities in Gujarat is that of extremely strong farmers' lobby groups (Bala, 2015).

Flat tariffs for electricity to pump groundwater and informal water markets

The decades since 1970s saw a spectacular increase in the use of electrical pumps to abstract groundwater. These increased by 585% during the period 1970-2001 to close to 350 000 pumps[4] (Shah et al., 2008). Initially, the Gujarat Electricity Board (GEB) charged farmers using electrical pumps based on their metered consumption of electricity. This scheme soon turned out to have certain disadvantages, such as high transaction costs and growing corruption linked to billing and metering. Moreover, the scheme was highly contested by farmers, who complained about the arbitrary nature of meter readers. Consequently, the GEB replaced it in 1988 by a scheme based on flat electricity tariffs linked to horsepower of pumps. As a result, the marginal cost of electricity consumption fell to zero and tube well owners were still not charged for the groundwater resource itself. This provided them with a strong incentive to sell groundwater to neighbours that did not possess their own wells. A dense informal groundwater market developed, with prices being pushed down by competition among the sellers. This greatly benefitted poor smallholders, who did not possess their own wells, leading to great access to groundwater for smallholder irrigation (Shah and Verma, 2008). As an increased number of farmers gained access to larger quantities of groundwater, agricultural productivity expanded.

Conversely, the flat tariff scheme negatively impacted farmers in the sense that they were obliged to pay for electricity consumption all year long, including during seasons where the use of water for irrigation was minor. For GEB, the flat tariff scheme resulted in declined metering and billing costs, but increasingly high costs of electricity subsidies, resulting from the rising electricity consumption (Shah and Verma, 2008). The flat tariff remained constant while consumption and actual costs rose. In 2000-01, the electricity subsidies made up as much as 56% of the fiscal deficit of Gujarat (Cullet, 2014).

The unsustainable increase in water withdrawals, the increasing consumption of electricity and the growing fiscal deficit were major drawbacks to the flat tariff scheme. GEB attempted to increase the flat tariff, but fell short of doing so because of strong opposition by farmers' lobby groups (Shah et al., 2008). As groundwater levels dropped, well owners invested in bigger pumps, aggravating the existing problems. The groundwater overdraft was a stark concern already in the mid-1980s, and groundwater depletion assumed the proportions of a crisis in certain areas of Gujarat during the 1990s (Shah et al., 2008) (Figure 11.1).

During the 1990s, GEB began limiting the number of hours of power supply per day.[5] However, unintentionally, this also impacted the power supply for domestic users, as

Figure 11.1. **Sharp rise in irrigation from groundwater in Gujarat, 1971-2001**

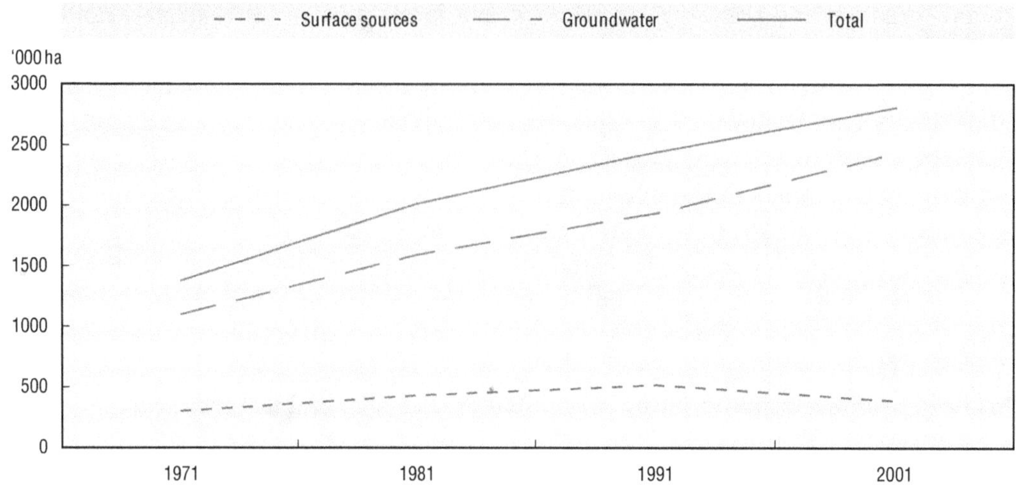

Source: Shah et al., 2008.

agricultural and domestic power was fed through the same system. As a result, the villages were left with weakened and unpredictable access to electricity (Shah et al., 2008; Shah and Verma, 2008)

The Jyotigram Scheme, 2003-06

Based on scientific evidence and recommendations made by the International Water Management Institute, GEB chose to combine the flat electricity tariffs with the introduction of rationing and real-time co-management of electricity and groundwater for agriculture. The new scheme was named "Jyotigram" – "the light of the village" (Shah et al., 2008). Launched during the period 2003-06, the scheme entailed separating the power supply for agricultural use from that for commercial and residential use, which required an investment of about USD 290 million (Shah et al., 2008). The total rewiring of Gujarat was a complicated process, yet by 2006, 90% of all 18 000 villages in the state were integrated in the scheme (Shah and Verma, 2008). With a functional parallel supply network put in place, it was possible to implement targeted rationing: the non-farming sectors were given access to 24 hours full-voltage, metered power supply seven days a week, while the farmers were provided with 8 hours full-voltage supply per day, at predictable times. The power supply to farmers remained highly subsidised, whereas the supply for non-agricultural use is charged based on metered consumption (Grönwall, 2014).

The Jyotigram scheme brought numerous advantages to the non-agricultural sectors in Gujarat which enjoy continuous access to power at full voltage. Farmers now have improved predictability and strength of power supply. This has allowed them to maintain their irrigation schedules so as to use labour more efficiently, conserve water, and save on pump maintenance costs. Furthermore, the Jyotigram scheme is estimated to have resulted in a 37% reduction in farm power use for tube wells from 2001 to 2006. This allowed for a decrease in aggregate farm power supply, thus a considerable improvement of GEB's financial viability. The aggregate farm power subsidy fell from USD 788 million in 2001-02 to USD 388 million in 2006-07 (Shah et al., 2008). From having had annual losses of between USD 119 and 550 million in 1999-2003, due to subsidy expenses, GEB ended up gaining a surplus of USD 50 million in 2006, and was considered to have performed the best power management in all of India (Shah et al., 2008).

Further, no decrease in agricultural yield has been observed in Gujarat (Rayfuse and Weifelt, 2013). During the seven years following the implementation of the scheme, agricultural GDP in the state rose by close to 10%, the highest in India, an increase that can be attributed to a broad range of factors (Planning Commission, 2012; CGIAR, 2012).[6]

While it is impossible to measure exactly the impact of the scheme on groundwater level (Shah et al., 2008), groundwater levels in the north of Gujarat have been rising by an average of four meters annually during recent years, compared to an annual drop of three meters per year before the launch of the scheme (Gupta, 2011). Another study indicates that the drop in water tables has at least slowed, albeit not everywhere across Gujarat (Narula et al., 2011).

However, there are some drawbacks to the scheme. For several farmers, the investment in a tube well was made viable by the fact that they could run it during 18-20 hours per day and sell groundwater on the informal market. The Jyotigram scheme obliged them to end this activity, thus reducing their income. The limitation to water pumping has also negatively impacted farmers who do not possess their own tube wells, as their access to groundwater now comes at a higher cost. Since well-owners can no longer pump unlimited amounts of water, the quantity of water being sold on the informal market is now more restricted, thus more expensive (Shah et al., 2008). Cash sales of pump irrigation water on informal markets have increased by 40-60% since the Jyotigram scheme was implemented, and several poorer farmers, not possessing their own tube wells, have been obliged to reduce their total area of irrigated land (Shah and Verma, 2008).

Lessons learned

The Jyotigram scheme, which has been replicated in at least seven other Indian states, illustrates how integrated policies for electricity and groundwater allocation can have mutual benefits for the conservation of both resources. This reflects the advantages of policy coherence across sectors that affect groundwater allocation (see Health Check #10, Part I). In a context where metered tariffs for electricity were difficult to enforce because of strong opposition, transaction costs and corruption, the combination of a bifurcated power supply system, flat tariffs and rationing appears to be a practical solution.

The scheme has created enhanced predictability in terms of quantity and quality of electricity access for both farmers and non-farmers, resulting in a significant decline in the power consumed by the agricultural sector and cost of related subsidies. As for groundwater, the Jyotigram scheme resulted in decreased consumption, allowing for depletion to slow down. Moreover, tube well owners have experienced declined risk in terms of pump maintenance costs and power shortage. The main drawback to the Jyotigram scheme is its implications for farmers that do not possess their own tube wells; additional policy measures are needed in order to improve their access to groundwater (Grönwall, 2014; CGIAR, 2012).

Notes

1. Regulatory measures could include grant of permits for sinking new bore wells, registration of existing bore well owners, registration of drilling agencies, restrictions on the depth and diameter of bore wells, restriction on purpose of use of groundwater, registration of new users in non-notified areas, adoption of rain harvesting, and penalty of offences (Kaushik, 2016).

2. However, the Groundwater Model Bill of 2011 has not formally replaced the 1970/2005 bill and there is no hierarchy between the two bills.
3. The number of well owners had grown to more than 1.044 million (Shah et al., 2008).
4. Still, there were slightly more diesel pump abstracting groundwater in Gujarat in 2001.
5. During the 1980s, the villages in Gujarat had access to 18-20 hours of three-phase electricity per day. This declined to just 10-12 hours per day by 2000 (Shah et al., 2008).
6. These include: the promotion of water-saving irrigation technology, mass based water harvesting and groundwater recharge, reform of agricultural marketing institutions, and a revitalised agricultural extension system (Gulati et al., 2009). Rapid adoption of new varieties of crops such as genetically modified cotton varieties, investments in rural roads, as well as favourable monsoons over the last decade, have also contributed to the strong increase in agricultural GDP (Planning Commission, 2012; Gulati et al., 2009).

References

Bala, R. (2015), "Policies intervention for groundwater governance in Gujarat and politics", *International Research Journal of Social Sciences*, Vol. 4/1, pp. 55-58, *www.isca.in/IJSS/Archive/v4/i1/9.ISCA-IRJSS-2014-230.pdf*.

Census India (2011), "Gujarat Profile", *http://censusindia.gov.in/2011census/censusinfodashboard/stock/profiles/en/IND024_Gujarat.pdf* (accessed 25 May 2016).

CGIAR (2012), "Getting to grips with India's groundwater", CGIAR, *www.cgiar.org/consortium-news/getting-to-grips-with-indias-groundwater/* (accessed 20 July 2016).

Cullet, P. (2014), "Groundwater law in India: Towards a framework ensuring equitable access and aquifer protection", *Journal of Environmental Law*, Vol. 26/1, pp. 55-81.

Desai, D. (2013), "Controversial Gujarat irrigation bill gets Governor's nod", *The Hindu*, *www.thehindu.com/news/national/controversial-gujarat-irrigation-bill-gets-governors-nod/article4562494.ece* (accessed 26 July 2016).

Grönwall, J. (2014), "Power to segregate: improving electricity access and reducing demand in rural India", Stockholm International Water Institute, Paper 23, *www.siwi.org/wp-content/uploads/2015/09/Power_to_Segregate.pdf* (accessed 9 August 2016).

Gupta, R. (2011), "The role of water technology in development: a case study of Gujarat, India", UN Water, *www.un.org/waterforlifedecade/green_economy_2011/ppt/04_10_2011_market_place_india_rajiv_kumar_gupta.pdf*.

Kaushik, Y.B. (2016), "Model bill for regulation of groundwater development", Central Groundwater Board of India, *http://mowr.gov.in/writereaddata/2-1-1_Model%20Bill%20to%20Regulate%20GW Development.pdf* (accessed 21 July 2016).

Narula, K. et al. (2011), "Addressing the water crisis in Gujarat, India", Columbia Water Center, Earth Institute, Columbia University, *http://water.columbia.edu/files/2011/11/Gujarat-WP.pdf* (accessed 21 July 2016).

Parekh, N. (2014), "Gujarat Irrigation and Drainage Act: Moving towards regulating groundwater use", CIPT Sandesh, issue 4, Columbia Water Center and Center for Intenational Projects Trust, *http://water.columbia.edu/files/2015/02/CIPT-Sandesh_Issue-4.pdf* (accessed 18 July 2016).

Planning Commission (2012), "Twelfth five year plan (2012-2017): Economic sectors, volume 2", Planning Commission of India, *http://planningcommission.gov.in/plans/planrel/12thplan/pdf/12fyp_vol2.pdf* (accessed 18 July 2016).

Planning Commission (2011), "Draft model bill for the conservation, protection and regulation of groundwater, 2011: Background and rationale", Planning Commission of India, *www.planningcommission.nic.in/aboutus/committee/wrkgrp12/wr/wg_back.pdf* (accessed 18 July 2016).

Rayfuse and Weifelt (eds.) (2013), *The Challenge of Food Security: International Policy and Regulatory Frameworks*, Edward Elgar Publishing Limited, Cheltenham, UK, and Northampton, US.

Shah, T. et al. (2008), "Groundwater governance through electricity supply management: assessing an innovative intervention in Gujarat, western India", *Agricultural Water Management* Vol. 95/11, pp. 1233-1242, *http://dx.doi.org/10.1016/j.agwat.2008.04.006*.

Shah, T. and S. Verma (2008), "Co-management of electricity and groundwater: an assessment of Gujarat's Jyotirgram scheme", *Economic and Political Weekly*, Vol. 43/7, pp. 59-66, *http://www.epw.in/journal/2008/07/special-articles/co-management-electricity-and-groundwater-assessment-gujarats*.

Water Governance Facility (2013), "Groundwater governance in India: Stumbling blocks for law and compliance", *Water Governance Facility Report* No. 3, SIWI, Stockholm.

Groundwater Allocation
Managing Growing Pressures on Quantity and Quality
© OECD 2017

PART II

Chapter 12

Flexibility in allocation through informal water trading in North China

This chapter explores how informal water trading has provided flexibility in groundwater allocation in North China. The case discusses how informal groundwater markets emerged as a result of well privatisation as well as some of the distributional considerations that arise from informal markets. The case also discusses the influence of pumping costs on trading activity and groundwater consumption.

Groundwater resources and use in North China

The population and economy in North China are highly dependent on groundwater resources. As of 2011, 35.5% of the total water supply in the region came from groundwater resources, whereas the share of groundwater in agricultural water supply amounted to about 70% (2004) (Wang et al., 2014). Close to 95% of China's 3.5 million tube wells are situated in the northern part of the country (Zhang et al., 2008; Mukherji and Shah, 2005).

One of the most important groundwater sources in North China is the extensive and complex aquifer system underlying the North China Plain, which is the leading agricultural area in China. The region supplies 61% of the country's wheat, 45% of its maize, 35% of cotton and 64% of peanuts (Yang et al., 2015). The North China Plain aquifer system consists of one shallow unconfined aquifer and three deep confined ones of different depths (Feng et al., 2013).

Extensive pumping has resulted in significant groundwater depletion

Despite the fact that North China produces 38% of the country's GDP and more than half the country's grain yield, the region has only 21% of the nation's water resources (Wang et al., 2016; Yang et al., 2015). Water access has declined as the groundwater levels in the North China Plain Aquifers have fallen dramatically, due to expanded irrigation and urbanisation (Changming et al., 2001). Over the period 1995-2004, the water table dropped in 48% of the villages in North China; and 8% were subject to severe overdraft, with the water table falling by more than 1.5 meter annually (Wang et al., 2007). The groundwater depletion in North China threatens long-term agricultural and industrial development in the region, and is expected to alter the balance of economic activity (Foster and Garduno, 2004).

Groundwater depletion in North China has also caused seawater intrusion and land subsidence. Land subsidence resulted in the collapse of more than 200 buildings already before 1995, and has created cones of depression under some cities (Foster and Garduno, 2004; Changming et al., 2001; Wang et al., 2014).

Regulation and enforcement constitute key challenges

As of 2002, Chinese Water Law states that all groundwater resources, including the right to use, sell and charge for groundwater, belong to the government. In practice, however, villages overlying aquifers have a *de facto* right to use the resources. Thus, groundwater entitlements are not related to land ownership or historic use entitlements, but primarily to ownership of wells (Wang et al., 2014).

The regulatory framework for groundwater management in China and its enforcement is generally weak (Mukherji and Shah, 2005). For example, the issuing of water extraction permits is often delayed and complicated. According to a survey carried out across China in 2004, only 10% of Chinese drillers surveyed held an extraction permit, despite this being nearly a universal requirement across the country (Wang et al., 2009). No abstraction charges or quantity limitations were imposed on well owners in any of the villages surveyed. Only 5% of community leaders surveyed believed that well drilling decisions required considerations of well spacing requirements (Wang et al., 2009).

Also at national level, groundwater management is constrained by limited resources. The resources devoted to groundwater management at ministerial level are considerably smaller than those for surface water management and flood control (Wang et al., 2009). Furthermore, there are no single management authorities for those aquifers that span jurisdictional boundaries. As a result, the co-ordination of users within aquifers that span across several regions is limited, and weakens the enforcement of governmental regulations (Mukherji and Shah, 2005). Groundwater governance and management is primarily carried out on a village level. Although localised management has advantages, it makes it challenging to implement programmes requiring collective action, such as universal water savings (Wang et al., 2014).

Nevertheless, some improvement in state governance of groundwater has been observed. Groundwater governance has changed from being highly fragmented to being more integrated and institutionalised. Agencies have gotten clearer responsibilities, and user participation in the form of water-user associations has been given significant consideration (Mukherji and Shah, 2005).

The privatisation of wells led to the emergence of groundwater markets

Before the agrarian reforms of the Deng administration in 1979, wells were collectively owned all over China (Mukherji and Shah, 2005). They were financed by collective earnings and resources from the township governments. The pumps were provided by state-run local agricultural inputs corporations or water resource bureaus. The village leaders made all decisions regarding timing and location for the wells, and the quantity of water that would be extracted per season. Farmers in the villages contributed their labour to the tube well construction and maintenance. The local, collective management of wells was based on a set of simple rules for groundwater allocation, and all individuals were provided with an equitable share of water (Zhang et al., 2008; Wang et al., 2007; 2014).

The economic and rural reforms of the late 1970s and 1980s aimed at accelerating growth in a number of sectors, notably agriculture. The economic reforms required local village governments to be fiscally more independent. As a result, several villages experienced considerable economic difficulties and were no longer able to invest in agriculture, including in the establishment and maintenance of collective wells. A coinciding fall in groundwater levels caused an additional decline in the number of functioning wells. Moreover, governmental regulations increasingly relaxed their restrictions on private activities and allowed for expanded freedom for individuals to invest in their own farms. The income and control rights of land were shifted from the collective to the individual household. Consequently, several farmers started sinking and operating their own wells (Wang et al., 2014). The number of private wells had been close to zero during the 1980s, but rose to approximately 40% in the 1990s. By 2004, 70% of the tube wells were privately owned (Zhang et al., 2008).

The economic reforms and the emergence of privately owned wells facilitated and encouraged the establishment of informal groundwater markets (Zhang et al., 2008; Easter and Huang, 2014). In 1994, groundwater markets only existed in 9% of the villages of North China, whereas by 2004, such markets had appeared in as much as 44% of the villages. In

1995, only water from 5% of tube wells was sold on the groundwater markets; this rate had risen to 18% by 2004 (Zhang et al., 2008).

Groundwater markets in China are localised and informal

Groundwater markets in North China are for the most part informal; transactions between sellers and buyers take place without legal contracts and sanctions. Nevertheless, some sort of formal regulation applies in 20-25% of the villages in North China, often in the form of a price ceiling. In other places, local authorities have influenced groundwater market activity by providing grants and loans for tube well construction, stimulating expanded market activity (Wang et al., 2014; 2016; Zhang et al., 2008).

Most markets operate within one single village. Only 6% of water-selling well owners sell to other villages than their own (Zhang et al., 2008). Contrary to the practice in many Southern Asian countries, in groundwater markets in North China water is sold at the same price regardless of the customer: only 7% of water sellers report that they charge different prices depending on the type of buyer (Zhang et al., 2008). Because the price of electricity in China is based on metered consumption, the depth from which water is pumped influences the price at which well owners sell groundwater on the market (Zhang et al., 2008).

The groundwater markets, access to and depletion of groundwater are strongly interlinked

The groundwater markets in North China provide a mechanism for the allocation of scarce water resources. Compared to a situation where wells had been privatised but no groundwater markets had emerged, markets create enhanced access to water for those who otherwise would struggle to access groundwater, such as poor, old and less educated farmers (Zhang et al., 2008; Wang et al., 2007; 2016). Research shows that 70% of sample households depend on groundwater for irrigation, whereas only 35% of these have their own wells. Some people still access groundwater through collective wells, but more than 20% depend on the informal markets for access to groundwater for irrigation (Zhang et al., 2008). The income of water-buying households is on average 61% of that of water-selling ones, and most of these would not be able to afford to invest in their own tube well. In addition to expanding groundwater access, the groundwater markets have become an important source of income for tube well owners (Wang et al., 2016).

Groundwater markets' implications for equitable access have been challenged by the steadily increasing groundwater depletion. As water levels have fallen, tube wells have been sunk to deeper levels, and the price of water has gone up due to increased electricity costs. Consequently, the access of some of the poorest farmers has been restrained (Zhang et al., 2008). Further, some scholars argue that the privatisation of wells and the emergence of informal groundwater markets have resulted in increased depletion, since this facilitates groundwater usage for a larger number of people. Thus, in the long run, in the absence of a limit on abstraction, markets may actually end up limiting access to water by increasing the scarcity of the resource (Song and Woo, 2008; Wang et al., 2016).

Conversely, there is empirical evidence showing that market actors respond to groundwater scarcity by reducing groundwater use, making water consumption more efficient and turning to crops that are less water-intensive (Song and Woo, 2008; Wang et al., 2006). When the water price increases because pumping costs go up, farmers seek to reduce their usage of groundwater while maintaining their crop production. This effect spreads in the groundwater markets. By creating a price signal, some argue that groundwater markets have encouraged efficiency of use, without harming production or income (Zhang et al., 2008; Wang et al., 2007).

Lessons learned

The privatisation of groundwater wells in China gave rise to informal water trading. Groundwater markets in North China constitute a means for reallocation of water among users (see Health Check #14, Part I). The groundwater markets have allowed for increased groundwater access for farmers that lack the means to install their own wells.

Increased groundwater scarcity tends to lead to expanded groundwater market activity, but also more efficient use of the water resources. Because electricity tariffs in China are set based on metered consumption, the depth from which groundwater is pumped determines the costs of operating a tube well. When pumping costs are higher, water sellers as well as buyers tend to optimise their groundwater consumption, at least in terms of their private use. Due to the informal market's responsiveness to price changes, some observers argue that the government should introduce a formal groundwater pricing mechanism, allowing for the recovery of the full costs of supply and reinforcing the price signal to reflect the scarcity of the resource (Wang et al., 2016).

References

Changming, L., Y. Jinjie and E. Kendy (2001), "Groundwater exploitation and its impact on the environment in the North China Plain", *Water International*, Vol. 23/2, pp. 265-272, International Water Resources Association.

Easter, K. and Q. Huang (eds.) (2014), "Water markets for the 21st century: What have we learned?", *Global Issues in Water Policy*, 11, Springer Science+Business Media, Dordrecht.

Feng et al. (2013), "Evaluation of groundwater depletion in North China using the Gravity Recovery and Climate Experiment (GRACE) data and ground-based measurements", *Water Resources Research*, Vol. 49, *http://dx.doi.org/10.1002/wrcr.20192*.

Foster, S. and H. Garduno (2004), "China: Towards sustainable groundwater resource use for irrigated agriculture on the North China Plain", *Sustainable Groundwater Management Lessons from Practice Case Profile Collection*, No. 8, The World Bank, Washington, DC, *http://documents.worldbank.org/curated/en/2004/12/12931846/china-towards-sustainable-groundwater-resource-use-irrigated-agriculture-north-china-plain* (accessed 11 July 2016).

Mukherji, A. and T. Shah (2005), "Groundwater socio-ecology and governance: a review of institutions and policies in selected countries", *Hydrogeology Journal*, Vol. 13(1), pp. 328-345.

Song, L. and W. Woo (eds.) (2008), "China's dilemma: Economic growth, the environment, and climate change", Anu E Press, Asia Pacific Press, Brookings Institution Press and Social Sciences Academic Press (China).

Wang et al. (2016), *Managing Water on China's Farms: Institutions, Policies and the Transformation of Irrigation under Scarcity*, Academic Press, Elsevier.

Wang et al. (2014), "Assessment of the development of groundwater market in rural China", in Easter, K.W. and Q. Huang (eds.), *Water Markets for the 21st Century: What Have We Learned?*, Global Issues in Water Policy 11, Springer Science+Business Media, Dordrecht 2014.

Wang et al. (2009), "Understanding the water crisis in northern china: What the government and the farmers are doing", *International Journal of Water Resources Development*, Vol. 25(1).

Wang, J. et al. (2007), "Agriculture and groundwater development in northern China: Trends, institutional responses, and policy options", Water Policy, Vol. 9/1, pp. 61-74, *http://dx.doi.org/10.2166/wp.2007.045*.

Yang et al. (2015), "Recharge and groundwater use in the North China Plain for six irrigated crops for an eleven year period", *PLoS One*, Vol. 10/1, *http://dx.doi.org/10.1371/journal.pone.0115269*.

Zhang et al. (2008), "Development of groundwater markets in China: A glimpse into progress to date", *World Development*, Vol. 36/4, pp. 706-726, *http://dx.doi.org/10.1016/j.worlddev.2007.04.012*.

Glossary

Abstraction: The capture, diversion, taking of water for any purpose including an environmental purpose.

Allocation regime: The combination of policies, laws and institutional arrangements (entitlements, licenses, permits, etc.) used to determine who is allowed to abstract water from a resource pool, how much may be taken and when, as well as how much must be returned (of what quality), and the conditions associated with the use of this water.

Aquifer: Hydraulic continuous body of porous geological structure containing groundwater.

Groundwater depletion: See "unsustainable use".

Groundwater development stress indicator (GDS): The ratio of groundwater abstraction for a given year to the mean annual groundwater recharge (including induced and artificial recharge) usually expressed as a percentage. It is a useful measure for the probability of the occurrence of negative side effects of groundwater depletion (Margat and van der Gun, 2013).

Groundwater system: A connected body of water located beneath the earth's surface in soil pore spaces and/or in the fractures of rock formations.

Intensive development of groundwater: Development of the resource to such an extent to significantly change the natural flow in the aquifer or aquifer system (Margat and van der Gun, 2013).

Storativity: The amount of water released per unit area of aquifer in response to per unit decline in hydraulic head (Freeze and Cherry 1979 in Huang, et al 2012).

Sustainable yield: Flux of groundwater that can be withdrawn from an aquifer without causing undesirable side effects, in particular without causing a permanent state of imbalance in the hydrological budget of an aquifer (Margat and van der Gun, 2013).

Transmissivity: The speed of lateral flow of groundwater (Saak and Peterson, 2007).

Unsustainable use (also referred to as "over drafting" or "groundwater depletion"): Groundwater use beyond recharge capacity.

Usufructuary rights: The right of use of a resource or property and the enjoyment of benefits from that use. These use rights may be subject to conditions, such as the "reasonable" or "beneficial" use doctrine and limited to a pre-determined duration.

Water entitlement: The entitlement to abstract and use water from a specified resource pool as defined in the relevant water plan or legislation. In some countries, this may be referred to as "water rights", "water users' rights", "water contracts", abstraction license or permit.

ORGANISATION FOR ECONOMIC CO-OPERATION AND DEVELOPMENT

The OECD is a unique forum where governments work together to address the economic, social and environmental challenges of globalisation. The OECD is also at the forefront of efforts to understand and to help governments respond to new developments and concerns, such as corporate governance, the information economy and the challenges of an ageing population. The Organisation provides a setting where governments can compare policy experiences, seek answers to common problems, identify good practice and work to co-ordinate domestic and international policies.

The OECD member countries are: Australia, Austria, Belgium, Canada, Chile, the Czech Republic, Denmark, Estonia, Finland, France, Germany, Greece, Hungary, Iceland, Ireland, Israel, Italy, Japan, Korea, Latvia, Luxembourg, Mexico, the Netherlands, New Zealand, Norway, Poland, Portugal, the Slovak Republic, Slovenia, Spain, Sweden, Switzerland, Turkey, the United Kingdom and the United States. The European Union takes part in the work of the OECD.

OECD Publishing disseminates widely the results of the Organisation's statistics gathering and research on economic, social and environmental issues, as well as the conventions, guidelines and standards agreed by its members.

OECD PUBLISHING, 2, rue André-Pascal, 75775 PARIS CEDEX 16
(97 2015 39 1 P) ISBN 978-92-64-28152-3 – 2017

www.ingramcontent.com/pod-product-compliance
Lightning Source LLC
Chambersburg PA
CBHW061222240926
51846CB00006B/1316